Association of Ukrainian g.
(AUS DAAD)
National Committee IAESTE Ukraine
(NC IAESTE-Ukraine)

Series: "Modern Mathematics for Engineers"

Tamara G. Stryzhak / Тамара Стрижак

Analytical Functions of Matrices

Аналитические функции от матрицы

$$e^{\mathrm{A}t} F\left(e^{-\mathrm{A}t} X \right) \equiv F(X)$$

$$-\infty < t < \infty$$

Tamara G. Stryzhak / Тамара Стрижак

ANALYTICAL FUNCTIONS OF MATRICES

АНАЛИТИЧЕСКИЕ ФУНКЦИИ ОТ МАТРИЦЫ

ibidem-Verlag
Stuttgart

Bibliografische Information der Deutschen Nationalbibliothek
Die Deutsche Nationalbibliothek verzeichnet diese Publikation in der Deutschen Nationalbibliografie; detaillierte bibliografische Daten sind im Internet über http://dnb.d-nb.de abrufbar.

Bibliographic information published by the Deutsche Nationalbibliothek
Die Deutsche Nationalbibliothek lists this publication in the Deutsche Nationalbibliografie; detailed bibliographic data are available in the Internet at http://dnb.d-nb.de.

∞

Gedruckt auf alterungsbeständigem, säurefreien Papier
Printed on acid-free paper

ISBN-13: 978-3-8382-0269-3

© *ibidem*-Verlag
Stuttgart 2011

Printed in Germany

Project

"Modern Mathematics for Engineers"

includes publishing the following works:

1. Difference equation with random coefficients
2. Stability of solutions of differential equations systems with random coefficients
3. Random values modeling
4. Optimal control synthesis
5. The Principle of reduction
6. New method of averaging
7. New determinant theory
8. Minimax criterion of stability
9. Numerical methods of stability research
10. Analytical functions from matrix
11. Frequently criteria of stability

This work describes functions of the square matrix. It gives general information about location and properties of functions of the matrix. It also discusses the problem of matrix eigenvalues, approximate methods of finding eigenvalues. It considers some questions connected with the matrix theory.

The "New Method of Averaging", which is connected with the matrix theory, is described in this work.

E-mail: stri@aer.ntu-kpi.kiev.ua
Website: www.iaeste.org.ua

"Modern Mathematics for Engineers"

The Author
*Prof. Tamara Stry*hak

Translator
Nataliya Sarycheva

IAESTE trainee students in 2010, that took part in the project *"Modern Mathematics for Engineers"* and in particular helped to prepare this work for publishing

Markus Stoller
Switzerland
Bern University of
Applied Sciences

Rachel McAdams
UK
University of
Warwick

Emma Woodham
UK
University of St.
Andrews

Simon
Konzett
Austria
University of
Vienna

Sonja
Remler
Germany
University of
Erlangen

Barbara
Peutler
Germany
Technical
University of
Augsburg

Michaela
Ziegler
Germany
University of
Munich

Table of Contents

1. General information about vectors and matrixes

Since vectors and matrixes are studied at all Universities we shall only present the information which is necessary for understanding this work.

Definition. Ordered set m of complex numbers $x_1, x_2, ..., x_m$ is called an   -measured vector and found

$$X = \begin{pmatrix} x_1 \\ x_2 \\ ... \\ x_m \end{pmatrix}, \quad X^* = (x_1, x_2, ..., x_m).$$

Vector X is called a column vector, X^* is called a row vector. Transfer from notation   to notation X^* is called vector transposition. We have

$$\left(X^*\right)^* = X.$$

Definition. Scalar multiplication product of two vectors

$$X = \begin{pmatrix} x_1 \\ x_2 \\ ... \\ x_m \end{pmatrix}, \quad Y = \begin{pmatrix} y_1 \\ y_2 \\ ... \\ y_m \end{pmatrix},$$

is number

$$Y^* X = X^* Y = (X, Y) = y_1 x_1 + y_2 x_2 + ... + y_m x_m.$$

$$(1)$$

Vectors X, Y are called mutually orthogonal if $Y^*X = 0$.

In order to characterize vector   as one number we shall introduce norm $\|X\|$ of the vector.

Norm $\|X\|$ is a real number. The norm of the vector can be derived in many ways. For this the following properties must be performed

1. $\|X\| \geq 0$. Если $\|X\| = 0$, то $X = 0$,

2. $\|\lambda X\| = |\lambda| \cdot \|X\|$,

3. $\|X + Y\| \leq \|X\| + \|Y\|$.

$$(2)$$

The norm of the vector is often found

1) $\|X\| = \max_i |x_i| \quad (i = 1, 2, ..., m)$,

2) $\|X\|_1 = |x_1| + |x_2| + ... + |x_m| = \sum_{i=1}^{m} |x_i|$,

3) $\|X\|_2 = \sqrt{X^*X} = \sqrt{|x_1|^2 + |x_2|^2 + ... + |x_m|^2}$.

$$(3)$$

All the norms of the vector are equivalent, i.e. for finding norms $\|X\|_\alpha$, $\|X\|_\beta$ of the vectors there are such vector constants c_1, c_2 that

$$c_1\|X\|_\alpha \leq \|X\|_\beta \leq c_2\|X\|_\alpha \quad (0 < c_1 \leq c_2).$$

We shall consider a set of vectors $X_1, X_2, ..., X_n$. These vectors are called linear independent if the equality

$$\alpha_1 X_1 + \alpha_2 X_2 + ... + \alpha_n X_n = 0$$

(4)

is true only at $\alpha_1 = 0, \alpha_2 = 0, ..., \alpha_n = 0$.

Let $n \leq m$ and vectors   have the expression via projections

$$X_1 = \begin{pmatrix} x_{11} \\ x_{21} \\ x_{31} \\ .... \\ x_{m1} \end{pmatrix}, \quad X_2 = \begin{pmatrix} 0 \\ x_{22} \\ x_{32} \\ .... \\ x_{m2} \end{pmatrix}, \quad X_3 = \begin{pmatrix} 0 \\ 0 \\ x_{33} \\ .... \\ x_{m3} \end{pmatrix}, ...,$$

$$X_n = \begin{pmatrix} 0 \\ 0 \\ ... \\ x_{nn} \\ .... \\ x_{mn} \end{pmatrix}.$$

(5)

If $x_{11} \neq 0, x_{22} \neq 0, x_{33} \neq 0, ..., x_{nn} \neq 0$, then vectors are linearly independent.

<u>Definition</u>. System m of linearly independent vectors in   - dimensional space is called the basis.

Theorem. So that м vectors

$$X_1 = \begin{pmatrix} x_{11} \\ x_{21} \\ ... \\ x_{m1} \end{pmatrix}, \quad X_2 = \begin{pmatrix} x_{12} \\ x_{22} \\ ... \\ x_{m2} \end{pmatrix}, \quad ..., \quad X_m = \begin{pmatrix} x_{1m} \\ x_{2m} \\ ... \\ x_{mm} \end{pmatrix}$$

(6)

make the basis, it is necessary and sufficient for the determiner, the elements of which are the principle of vectors $X_1, X_2, ..., X_m$ which do not vanish.

$$\Delta \equiv \begin{vmatrix} x_{11} & x_{12} & ... & x_{1m} \\ x_{21} & x_{22} & ... & x_{2m} \\ ... & ... & ... & ... \\ x_{m1} & x_{m2} & ... & x_{mm} \end{vmatrix} \neq 0.$$

(7)

Definition. Vector X_0 is called a linear combination of vectors $X_1, X_2, ..., X_m$, if there are such numbers as $\alpha_1, \alpha_2, ..., \alpha_m$, and the equality is true

$$X_0 = \alpha_1 X_1 + \alpha_2 X_2 + ... + \alpha_m X_m.$$

(8)

Vector $\alpha_k X_k$ is called a projection of vector onto vector X_k.

Definition. V is a nonempty set of vectors from R^m. Set is called subspace from R^m if the condition $X \in V$, $Y \in V$ should be $\alpha X + \beta Y \in V$, where α, β are complex numbers.

Definition. Subspace V formed by vectors

$$X = \alpha_1 X_1 + \alpha_2 X_2 + ... + \alpha_s X_s \quad (s \leq m)$$

is called subspace created by vectors $X_1, X_2, ..., X_s$. The biggest number r of linearly independent vectors in the vector system $X_1, X_2, ..., X_s$ is called its rank and defined

$$r = rank(X_1, X_2, ..., X_s), \text{ или } r = rang(X_1, X_2, ..., X_s).$$

Computing the rank of the vector system is a complicated task at large values m. As a rule the vector system has form (5) or vector orthogonalization is used.

Definition. Equivalent transformations of the vector system are:

1) transposition of vectors;
2) multiplying vectors by numbers different from zero;
3) addition of other vectors multiplied by random numbers to one vector.

Theorem. During equivalent transformations of the vector system its rank is not changed.

In order to calculate the rank of the vector system, equivalent transformations should have the form when the rank can be easily calculated.

Some data from the matrix theory.

Definition. Matrix of $m \times n$ size is a rectangular table of numbers

$$A = \begin{pmatrix} a_{11} & a_{12} & \ldots & a_{1n} \\ a_{21} & a_{22} & \ldots & a_{2n} \\ \ldots & \ldots & \ldots & \ldots \\ a_{m1} & a_{m2} & \ldots & a_{mn} \end{pmatrix}.$$

We write $\dim A = m \times n$. Vector-row

$$b = \begin{pmatrix} a_{11} & a_{12} & \ldots & a_{1n} \end{pmatrix}$$

is a matrix of $1 \times n$ size. Column vector

$$a = \begin{pmatrix} a_{11} \\ a_{21} \\ \ldots \\ a_{m1} \end{pmatrix}$$

is matrix of $m \times 1$ size.

Horizontal rows of numbers are called matrix rows; vertical rows of numbers are called columns.

If the number of matrix rows is equal to the number of matrix columns the matrix is called square. The number of rows is called the matrix order. The row of numbers $a_{11}, a_{22}, \ldots, a_{mn}$ is called the main diagonal of the matrix.

Definition. Square matrix is called diagonal if all the elements located out of the main diagonal are equal to zero

Diagonal matrix is called a unit matrix if all the elements located on the main diagonal are equal to one. The unit matrix is denoted by E

$$E = \begin{pmatrix} 1 & 0 & \dots & 0 \\ 0 & 1 & \dots & 0 \\ \dots & \dots & \dots & \dots \\ 0 & 0 & \dots & 1 \end{pmatrix}.$$

Definition. Matrix transposition is substitution of matrix lines by matrix columns with preserving the order of their sequence. Matrix transposition is denoted by an asterisk

$$A = \begin{pmatrix} a_{11} & a_{12} & \dots & a_{1n} \\ a_{21} & a_{22} & \dots & a_{2n} \\ \dots & \dots & \dots & \dots \\ a_{m1} & a_{m2} & \dots & a_{mn} \end{pmatrix}, \quad A^* = \begin{pmatrix} a_{11} & a_{21} & \dots & a_{m1} \\ a_{12} & a_{22} & \dots & a_{m2} \\ \dots & \dots & \dots & \dots \\ a_{1n} & a_{2n} & \dots & a_{mn} \end{pmatrix}.$$

We shall consider two matrixes: matrix A of $m \times n$ size and matrix B of $n \times l$ size

$$A = \begin{pmatrix} a_{11} & a_{12} & \dots & a_{1n} \\ a_{21} & a_{22} & \dots & a_{2n} \\ \dots & \dots & \dots & \dots \\ a_{m1} & a_{m2} & \dots & a_{mn} \end{pmatrix}, \quad B = \begin{pmatrix} b_{11} & b_{12} & \dots & b_{1l} \\ b_{21} & b_{22} & \dots & b_{2l} \\ \dots & \dots & \dots & \dots \\ b_{n1} & b_{n2} & \dots & b_{nl} \end{pmatrix}.$$

Definition. Product of matrixes AB is matrix $C = AB$ of $m \times l$ size

$$C = \begin{pmatrix} c_{11} & c_{12} & \cdots & c_{1l} \\ c_{21} & c_{22} & \cdots & c_{2l} \\ \cdots & \cdots & \cdots & \cdots \\ c_{m1} & c_{m2} & \cdots & c_{ml} \end{pmatrix}$$

With the element found by formulas

$$c_{ks} = \sum_{i=1}^{n} a_{ki} b_{is} \qquad (k = 1,2,...,m;\ s = 1,2,...,l).$$

(9)

Some properties of the matrix multiplication:

1. $\alpha(AB) = (\alpha A) \cdot B = A \cdot (\alpha B)$.
2. $(A + B)C = AC + BC$.
3. $C(A + B) = CA + CB$.
4. $ABC = A \cdot (BC) = (AB) \cdot C$.
5. $(AB)^* = B^* A^*$.
6. $AE = A, \quad EA = A$.

In the general case $AB \neq BA$.

Definition. If $A \cdot B = E$, $\dim A = m \times m$, $\dim B = m \times m$, then matrix B is called inverse to matrix A and defined $B = A^{-1}$.

If square matrix has inverse matrix A^{-1}, then it is called non-specific, $\det A \neq 0$ and the matrix itself can have a view

$$A^{-1} = \frac{1}{\Delta} \begin{pmatrix} A_{11} & A_{21} & ... & A_{m1} \\ A_{12} & A_{22} & ... & A_{m2} \\ ... & ... & ... & ... \\ A_{1m} & A_{2m} & ... & A_{mm} \end{pmatrix},$$

(10)

$$\Delta = \det \begin{vmatrix} a_{11} & a_{12} & ... & a_{1m} \\ a_{21} & a_{22} & ... & a_{2m} \\ ... & ... & ... & ... \\ a_{m1} & a_{m2} & ... & a_{mm} \end{vmatrix}.$$

Here Δ is determiner of matrixes A, A_{ks} is an algebraic additions of elements a_{ks} of determiner Δ. We shall note that

$$AA^{-1} = E, \quad A^{-1}A = E, \quad (AB)^{-1} = B^{-1}A^{-1}.$$

2. Norm of the matrix

A is a square matrix of m-order. We find some norm $\|X\|$ of vector X

$$X = \begin{pmatrix} x_1 \\ x_2 \\ \dots \\ x_m \end{pmatrix}.$$

Norm of matrix   is number $\max\limits_{\|x\| \leq 1} \|AX\| = \|A\|$.

$$(11)$$

Finding norm $\|A\|$ brings to inequality

$$\|AX\| \leq \|A\| \cdot \|X\|, \tag{12}$$

which is true at any vector   . The following properties of the norm of the matrix come out of formulas (11), (12)

1. $\|A\| \geq 0$. Если $\|A\| = 0$, то $A = 0$;
2. $\|\alpha A\| = |\alpha| \cdot \|A\|$;
3. $\|A + B\| \leq \|A\| + \|B\|$;
4. $\|AB\| \leq \|A\| \cdot \|B\|$.

If a_{ks} $(k, s = 1, 2, \dots, m)$ is a random element of matrix   , then

$$|a_{ks}| \leq \|A\| \qquad\qquad . \tag{14}$$

Thus the norm of the determiner matrix T is agreed with the definition of the norm of the vector.

If $\|X\| = \max_i |x_i| \quad (i = 1,2,...,m)$, then the norm of

matrix A

$$A = \begin{pmatrix} a_{11} & a_{12} & ... & a_{1m} \\ a_{21} & a_{22} & ... & a_{2m} \\ ... & ... & ... & ... \\ a_{m1} & a_{m2} & ... & a_{mm} \end{pmatrix}$$

is found by formula

$$\|A\| = \max_k \sum_{s=1}^m |a_{ks}|.$$

If $\|X\|_1 = |x_1| + |x_2| + ... + |x_m|$, then

$$\|A\|_1 = \max_s \sum_{k=1}^m |a_{ks}|.$$

If $\|X\|_2 = \sqrt{\sum_{k=1}^m |x_k|^2}$, then $\|A\|_2 = \sqrt{\lambda}$, where λ is

the biggest value of quadratic function

$f(x) = \sum_{k=1}^m (a_{k1}x_1 + a_{k2}x_2 + ... + a_{km}x_m)^2$. If we introduce

asymptotic matrix

$$B = \|b_{ij}\|_1^m, \quad b_{ij} = \sum_{k=1}^m a_{ki}a_{kj} \quad (i, j = 1,...,m),$$

then λ is the biggest eigenvalue of matrix B. It is assumed that $B = A^*A$.

It is difficult to calculate Euclidean norm $\|A\|_2$ and in calculations it is often replaced by another matrix norm

$$\|A\|_3 = \sqrt{\sum_{k,s=1}^m |a_{ks}|^2}, \quad \|A\|_3 \geq \|A\|_2.$$

Example. We find matrix norm

$$A = \begin{pmatrix} 1 & 2 \\ 1 & -2 \end{pmatrix}$$

with different definitions of the vector norm.

$$\|A\| = \max\{|1| + |2|, |1| + |-2|\} = 3;$$

$$\|A\|_1 = \max\{|1| + |1|, |2| + |-2|\} = 4;$$

$$\|A\|_2 = \max_{x_1^2 + x_2^2 \leq 1} \sqrt{(x_1 + 2x_2)^2 + (x_1 - 2x_2)^2} = \sqrt{8};$$

$$\|A\|_3 = \sqrt{1^2 + 2^2 + 1^2 + (-2)^2} = \sqrt{10}.$$

If elements of matrix A are small to the modulus then, while finding matrix norm $\|A\|_3$, mistake is possible because of round-off errors.

We shall consider matrix series for convergence as an application of the concept of matrix norm.

When $A_0, A_1, A_2, ..., A_k, ...$ are matrixes of $m \times m$-order expression

$$A_0 + A_1 + A_2 + ... + A_k + ... \tag{15}$$

is called a matrix series. Matrix series is equal to set m^2 of usual scalar series. Matrix series (15) converges if there is limit

$$A = \lim_{N \to \infty} \sum_{k=0}^{N} A_k . \tag{16}$$

In its turn matrix series (16) converges absolutely if series from norm series (16) converges too

$$\|A_0\| + \|A_1\| + \|A_2\| + ... + \|A_k\| + ... \tag{17}$$

Example. We shall consider matrix series

$$S = E + A + A^2 + A^3 + ... + A^k + ... \tag{18}$$

This series converges absolutely if scalar series converges too

$$\|E\| + \|A\| + \|A^2\| + \|A^3\| + ... + \|A^k\| + ... \le 1 + \|A\| + \|A\|^2 + \|A\|^3 + ... + \|A\|^k + ...$$

Consequently matrix series (18) converges absolutely at $\|A\| < 1$. We shall multiply series (18) by matrix $E - A$ and receive equality

$$S(E - A) = E + A + A^2 + ... + A^k - A - A^2 - ... - A^{k+1} - ... = E .$$

Consequently equality is true

$$E + A + A^2 + A^3 + ... + A^k + ... = (E - A)^{-1}, \qquad . \tag{19}$$

This equality can be applied for solving the system of linear equations

$$X = B + AX, \quad \|A\| < 1, \tag{20}$$

by method of subsequent approximations

$$X_0 = 0, \quad X_{n+1} = B + AX_n \quad (n = 0,1,2,...); \quad X = \lim_{n \to \infty} X_n. \tag{21}$$

Vector equation (20) enables to find

$$, \quad X_1 = B, \quad X_2 = B + AB,$$

$$X_3 = B + AB + A^2B, \ ...$$

$$X_n = B + AB + A^2B + ... + A^{n-1}B,$$

$$X = \left(E + A + A^2 + ... + A^k + ...\right)B$$

i.e. $X = (E - A)B$.

Method (21) is convenient because possible errors while finding X_n do not influence the final result.

3. Eigenvectors and eigenvalues of matrix

We shall consider square matrix A of m-order.

Definition. Zero vector X is called eigenvector of matrix , if there is such number λ that

$$AX = \lambda X. \tag{22}$$

Number is called eigenvalue of matrix which corresponds to eigenvector . System of equations (22) can be written

$$AX = \lambda EX, \quad (\lambda E - A)X = 0, \quad X \neq 0.$$

(23)

Eigenvector X is a solution of the homogeneous system of linear algebraic equations. To make zero solution of equation system (23) possible it is necessary and sufficient when the determiner of the system is equal to zero, i.e.

$$\det(E\lambda - A) = \lambda^m + b_1 \lambda^{m-1} + \ldots + b_m = 0.$$

(24)

Equation (24) is called characteristic and its roots are eigenvalues of matrix A. In the general case eigenvalues of the matrix and projections of eigenvectors are complex numbers.

As follows from equation (24), the determiner of matrix is equal to the product of eigenvalues of matrix

$$\lambda_1 \lambda_2 \ldots \lambda_m = \det A. \quad (25)$$

Definition. Set of all eigenvalues of matrix is called thebspectrum of matrix and denoted *Sr*A.

Example. We shall find eigenvalues and eigenvectors of the matrix

$$A = \begin{pmatrix} 1 & 4 \\ 1 & 1 \end{pmatrix}.$$

We shall write characteristic equation

$$\det(\lambda E - A) = \begin{vmatrix} \lambda - 1 & -4 \\ -1 & \lambda - 1 \end{vmatrix} = \lambda^2 - 2\lambda - 3 = 0,$$

which has roots $\lambda_1 = -1$, $\lambda_2 = 3$.

At $\lambda = -1$, system of equations (23) looks like

$$\begin{cases} -2x_1 - 4x_2 = 0, \\ -x_1 - 2x_2 = 0, \end{cases} \quad X_1 = \begin{pmatrix} 2 \\ -1 \end{pmatrix}.$$

At $\lambda = 3$ we shall receive system of equations

$$\begin{cases} 2x_1 - 4x_2 = 0, \\ -x_1 + 2x_2 = 0, \end{cases} \quad X_2 = \begin{pmatrix} 2 \\ 1 \end{pmatrix}.$$

We shall check the validity of equality (22)

$$AX_1 = \begin{pmatrix} 1 & 4 \\ 1 & 1 \end{pmatrix} \begin{pmatrix} 2 \\ -1 \end{pmatrix} = \begin{pmatrix} -2 \\ 1 \end{pmatrix} = -1 \cdot X_1,$$

$$AX_2 = \begin{pmatrix} 1 & 4 \\ 1 & 1 \end{pmatrix} \begin{pmatrix} 2 \\ 1 \end{pmatrix} = \begin{pmatrix} 6 \\ 3 \end{pmatrix} = 3 \cdot X_2.$$

The spectrum of matrix A makes a set

$$SrA = \{-1; 3\}.$$

We shall demonstrate some properties of eigenvectors.

Theorem. If all eigenvalues of a matrix are different, then all eigenvalues of the matrix are linearly independent.

Definition. Square matrixes , B are called similar if there is some non-singular matrix T that

$$B = T^{-1}AT. \tag{26}$$

Theorem. If A, B are similar matrixes then they have identical eigenvalues.

Theorem. If matrix   of m-order has different eigenvalues, then matrix   is similar to diagonal matrix

$$\Lambda = \begin{pmatrix} \lambda_1 & 0 & ... & 0 \\ 0 & \lambda_2 & ... & 0 \\ ... & ... & ... & ... \\ 0 & 0 & ... & \lambda_m \end{pmatrix}. \tag{27}$$

We shall take as matrix T a matrix the columns of which are eigenvalues of matrix   . Here we receive equality

$$\Lambda = T^{-1}AT$$

or $T\Lambda = AT$. This matrix equality is divided into vector equalities

$$\lambda_k X_k = AX_k \quad (k = 1,2,...,m).$$

Definition. Matrix Λ (27) is called canonical form Jordan for matrix   .

Definition. If matrix   is similar to diagonal matrix (27), then it is said that matrix   has a simple structure. If matrix   has different eigenvalues then they have a simple structure. ***In some cases matrix with multiple eigenvalues can have a simple structure.***

Example. For matrix A we shall write characteristic equation

$$A = \begin{pmatrix} 1 & -1 \\ 1 & 3 \end{pmatrix},$$

$$\det(\lambda E - A) = \begin{vmatrix} \lambda - 1 & 1 \\ -1 & \lambda - 3 \end{vmatrix} = \lambda^2 - 4\lambda + 4 = 0.$$

This equation has 2-multiple root $\lambda = 2$. As follows from equation

$$\lambda X = AX, \quad (\lambda E - A)X = 0$$

we shall receive system of equations for projections of vector X

$$\begin{cases} (\lambda - 1)x_1 + x_2 = 0, \\ -x_1 + (\lambda - 3)x_2 = 0. \end{cases}$$

We find zero solution $x_1 = 1$, $x_2 = -1$. Matrix has only one eigenvector

$$X = \begin{pmatrix} 1 \\ -1 \end{pmatrix}.$$

Supposing that

$$T = \begin{pmatrix} -1 & 1 \\ 0 & -1 \end{pmatrix}, \quad T^{-1} = \begin{pmatrix} -1 & 1 \\ 0 & -1 \end{pmatrix},$$

we find

$$T^{-1}AT = \begin{pmatrix} 2 & 0 \\ 1 & 2 \end{pmatrix}.$$

The last matrix has Jordan form in the case of multiple eigenvalues.

Let matrix $T = \begin{pmatrix} -1 & 1 \\ 0 & -1 \end{pmatrix}$ have eigenvalues $\lambda_1, \lambda_2, ..., \lambda_m$, which correspond to eigenvectors $X_1, X_2, ..., X_m$. Supposing these vectors are linearly independent we shall find inverse matrix T^{-1}

$$T = (X_1 X_2 ... X_m), \quad T^{-1} = \begin{pmatrix} Y_1 \\ Y_2 \\ ... \\ Y_m \end{pmatrix}.$$

The rows of matrix   are denoted $Y_1, Y_2, ..., Y_m$. Here equalities $T^{-1}T = E$ are true, or

$$Y_k X_s = \delta_{ks} = \begin{cases} 1 & npu \ k = s \\ 0 & npu \ k \neq s \end{cases} \quad (k, s = 1, 2, ..., m).$$

Here the theorem is true.

Theorem. If matrix A has m-linearly eigenvectors   , then it can look like

$$A = \lambda_1 X_1 Y_1 + \lambda_2 X_2 Y_2 + ... + \lambda_m X_m Y_m.$$

To prove it we shall multiply matrix   by vector X_k on the right

$$A X_k = \sum_{s=1}^{m} \lambda_s X_s (Y_s X_k) = \lambda_k X_k,$$

which proves the theorem.

Definition. Spectral radius $\rho(A)$ of matrix A is the biggest modulus of its eigenvalues.

$$\rho(A) = \max_{j}\left\{\left|\lambda_j\right|\right\}. \tag{28}$$

The spectral radius can be found by formula

$$\rho(A) = \lim_{k \to +\infty} \sqrt[k]{\left\|A^k\right\|}. \tag{29}$$

To find the spectral radius by formula (27) we offered a calculating formula which can be easily realized on the computer.

1. We create a sequence of matrixes and their norms

$$\sigma_1 = \|A\|, \quad A_1 = A\sigma_1^{-1};$$

$$\sigma_{n+1} = \|A_n A_n\|, \quad A_{n+1} = A_n A_n \sigma_{n+1}^{-1}.$$

All matrixes A_k $(k = 1,2,3,...)$ have a unit.

2. Spectral radius $\rho(A)$ is found by formula

$$\ln\rho(A) = \lim_{N \to +\infty}\left(\ln\sigma_1 + \frac{1}{2}\ln\sigma_2 + \frac{1}{4}\ln\sigma_3 + ... + 2^{-N}\ln\sigma_{N+1}\right)$$

.

3. The calculating process stops if the following equality if performed

$$\left|2^{-N}\ln\sigma_{N+1}\right| < \delta,$$

where σ is a given accuracy.

We shall estimate the spectral radius. From equality (22) we find equality

$$\|AX\| = \|\lambda X\|,$$

and then we find inequality

$$|\lambda| \leq \|A\|, \quad \rho(A) \leq \|A\|. \tag{30}$$

At any finding of the norm of matrix $\|A\|$, agreed with the vector norm, eigenvalues of matrix A are smaller than $\|A\|$ to modulo.

4. Projectors

To understand functions of matrixes it is required to know projections of matrixes. Here is some information about projectors in short.

Definition. Matrix P is called idempotent of a projector if the following equality is true

$$PP = P. \tag{31}$$

In a particular case zero and unit matrixes are projections.

Theorem. If matrix is a projector, then matrix $P_1 = E - P$ is a projector too.

Proof. We shall find product

$$P_1 P_1 = (E - P)(E - P) = E - 2P + PP = E - 2P + P = E - P = P_1.$$

As follows from equality $P_1 P_1 = P_1$, matrix P_1 is also a projector.

The term "projector" can be explained as follows. Let be a matrix of linear transformation–projecting of a vector on some sub-space. As the repeated projecting of a vector is equivalent to one-time projecting, then

.

Let vectors $X_1, X_2, ..., X_m$ create a basis in space R^m. We shall expand random vector $Y \in R^m$ by basis

$$Y = \alpha_1 X_1 + \alpha_2 X_2 + ... + \alpha_m X_m. \tag{32}$$

In order to find coefficients $\alpha_1, \alpha_2, ..., \alpha_m$ we shall receive system of linear algebraic equations

$$y_1 = \alpha_1 x_{11} + \alpha_2 x_{12} + ... + \alpha_m x_{1m},$$
$$y_2 = \alpha_1 x_{21} + \alpha_2 x_{22} + ... + \alpha_m x_{2m},$$
$$\dots\dots\dots\dots\dots\dots\dots\dots\dots\dots\dots\dots\dots\dots\dots,$$
$$y_m = \alpha_1 x_{m1} + \alpha_2 x_{m2} + ... + \alpha_m x_{mm},$$

$$X_k \equiv \begin{pmatrix} x_{1k} \\ x_{2k} \\ ... \\ x_{mk} \end{pmatrix},$$

which can be written like vector

$$Y = TA; \quad T = \begin{pmatrix} x_{11} & x_{12} & ... & x_{1m} \\ x_{21} & x_{22} & ... & x_{2m} \\ ... & ... & ... & ... \\ x_{m1} & x_{m2} & ... & x_{mm} \end{pmatrix}, \quad A = \begin{pmatrix} \alpha_1 \\ \alpha_2 \\ ... \\ \alpha_m \end{pmatrix}.$$

Using inverse matrix T, we find vector A

$$T^{-1} Y = A. \tag{33}$$

Rows of matrix T^{-1} are denoted $Z_1, Z_2, ..., Z_m$

$$T^{-1} = \begin{pmatrix} Z_1 \\ Z_2 \\ ... \\ Z_m \end{pmatrix}.$$

Since $T^{-1}T = E$, we find equality

$$Z_k X_s = \delta_{ks} = \begin{cases} 1, & \text{если } k = s, \\ 0, & \text{если } k \neq s; \end{cases} \quad k, s = 1, 2, ..., m. \tag{34}$$

According to formula (33) we find coefficients α_k

$$\alpha_k = Z_k Y \quad (k = 1,2,...,m).$$

Finally we find expansion (32) by vector basis

$$Y = X_1 Z_1 \cdot Y + X_2 Z_2 Y + ... + X_m Z_m Y .$$

$$(35)$$

Matrixes

$$P_k = X_k Z_k$$

are projectors in the considered basis.

Expansion (32) can be written like

$$Y = P_1 Y + P_2 Y + ... + P_m Y ,$$ (36)

which is followed by equality

$$P_1 + P_2 + ... + P_m = \mathrm{E}.$$ (37)

We shall find product of different projectors

$$P_k P_s = (X_k Z_k)(X_s Z_s) = X_k (Z_k X_s) Z_s = X_k \delta_{ks} Z_s$$
$$(k,s = 1,2,...,m).$$

Finally we find equalities

$$P_k P_k = P_k, \quad P_k P_s = 0 \quad (k \neq s, \ k,s = 1,2,...,m).$$

$$(38)$$

Example. In space R^2 we choose basis

$$X_1 = \begin{pmatrix} 1 \\ 2 \end{pmatrix}, \quad X_2 = \begin{pmatrix} 1 \\ 3 \end{pmatrix}.$$

We find matrixes

$$T = \begin{pmatrix} 1 & 1 \\ 2 & 3 \end{pmatrix}, \quad T^{-1} = \begin{pmatrix} 3 & -1 \\ -2 & 1 \end{pmatrix}$$

and projections on basis vectors

$$P_1 = \begin{pmatrix} 1 \\ 2 \end{pmatrix}(3 \quad -1) = \begin{pmatrix} 3 & -1 \\ 6 & -2 \end{pmatrix},$$

$$P_2 = \begin{pmatrix} 1 \\ 3 \end{pmatrix}(-2 \quad 1) = \begin{pmatrix} -2 & 1 \\ -6 & 3 \end{pmatrix}.$$

It is easy to check execution of properties (36), (38)

$$P_1 P_1 = P_1, \quad P_2 P_2 = P_2, \quad P_1 P_2 = 0, \quad P_2 P_1 = 0, \quad P_1 + P_2 = E.$$

We expand, for example, vector $Y = \begin{pmatrix} 1 \\ 1 \end{pmatrix}$ by basis

$$Y = P_1 Y + P_2 Y, \quad \begin{pmatrix} 1 \\ 1 \end{pmatrix} = \begin{pmatrix} 3 & -1 \\ 6 & -2 \end{pmatrix}\begin{pmatrix} 1 \\ 1 \end{pmatrix} + \begin{pmatrix} -2 & 1 \\ -6 & 3 \end{pmatrix}\begin{pmatrix} 1 \\ 1 \end{pmatrix}.$$

We have expansion

$$\begin{pmatrix} 1 \\ 1 \end{pmatrix} = 2\begin{pmatrix} 1 \\ 2 \end{pmatrix} - 1 \cdot \begin{pmatrix} 1 \\ 3 \end{pmatrix}.$$

Example. Basis in R^3 creates vectors

$$X_1 = \begin{pmatrix} 1 \\ 1 \\ 2 \end{pmatrix}, \quad X_2 = \begin{pmatrix} 2 \\ 3 \\ 5 \end{pmatrix}, \quad X_3 = \begin{pmatrix} 3 \\ 4 \\ 6 \end{pmatrix}.$$

We make matrixes T, T^{-1}

$$T = \begin{pmatrix} 1 & 2 & 3 \\ 1 & 3 & 4 \\ 2 & 5 & 6 \end{pmatrix}, \quad T^{-1} = \begin{pmatrix} 2 & -3 & 1 \\ -1 & 0 & 1 \\ 1 & 1 & -1 \end{pmatrix}.$$

We find projectors for expansion by basis

$$P_1 = \begin{pmatrix} 1 \\ 1 \\ 2 \end{pmatrix} \begin{pmatrix} 2 & -3 & 1 \end{pmatrix} = \begin{pmatrix} 2 & -3 & 1 \\ 2 & -3 & 1 \\ 4 & -6 & 2 \end{pmatrix};$$

$$P_2 = \begin{pmatrix} 2 \\ 3 \\ 5 \end{pmatrix} \begin{pmatrix} -2 & 0 & 1 \end{pmatrix} = \begin{pmatrix} -4 & 0 & 2 \\ -6 & 0 & 3 \\ -10 & 0 & 5 \end{pmatrix};$$

$$P_3 = \begin{pmatrix} 3 \\ 4 \\ 6 \end{pmatrix} \begin{pmatrix} 1 & 1 & -1 \end{pmatrix} = \begin{pmatrix} 3 & 3 & -3 \\ 4 & 4 & -4 \\ 6 & 6 & -6 \end{pmatrix}.$$

It is easy to check properties of projections

$$P_k P_s = P_k \delta_{ks} \quad (k, s = 1, 2, 3), \quad P_1 + P_2 + P_3 = E.$$

We find expansion of vector $Y = \begin{pmatrix} 2 \\ 3 \\ 1 \end{pmatrix}$ by basis

$$P_1 Y = \begin{pmatrix} -4 \\ -4 \\ -8 \end{pmatrix} = -4X_1, \quad P_2 Y = \begin{pmatrix} -6 \\ -9 \\ -15 \end{pmatrix} = -3X_2,$$

$$P_3 Y = \begin{pmatrix} 12 \\ 16 \\ 24 \end{pmatrix} = 4X_3.$$

From here we find expansion

$$Y = -4X_1 - 3X_2 + 4X_3.$$

If projectors are known, then expansion of the vector by basis is reduced to multiplication of the vector by the projectors.

Remark_. Let P be some projector, i.e. $PP = P$. Then matrix

$$B = 2P - E$$

is a root from a unit matrix, i.e.

$$B = \sqrt{E}, \quad B^2 = E.$$

We have equality

$$B \cdot B = (2P - E)(2P - E) = 4B^2 - 4P + E = E.$$

Inversely, if $B = \sqrt{E}$ is any root from a unit matrix, then matrix

$$P = \frac{1}{2}(E + \sqrt{E})$$

is a projector. In fact, we have equality

$$PP = \frac{1}{4}\left(E + 2E\sqrt{E} + E\right) = \frac{1}{2}\left(E + \sqrt{E}\right) = P.$$

Example. Matrix

$$P = \begin{pmatrix} 3 & -1 \\ 6 & -2 \end{pmatrix}$$

is a projector. At this matrix

$$B = 2P - E = \begin{pmatrix} 5 & -2 \\ 12 & -5 \end{pmatrix}$$

is a root from a unit matrix.

Let matrix A have linearly independent eigenvectors $X_1, X_2, ..., X_m$. If the basis is formed by eigenvectors of matrix , then corresponding projectors are called projectors of matrix .

If matrix has eigenvalues $\lambda_1, \lambda_2, ..., \lambda_m$, then we can restore the matrix itself by formula

$$A = \lambda_1 P_1 + \lambda_2 P_2 + ... + \lambda_m P_m. \tag{39}$$

In fact we find product

$$AX_k = \left(\lambda_1 P_1 + \lambda_2 P_2 + ... + \lambda_m P_m\right)X_k = \lambda_k X_k,$$

since $P_s X_k = 0$ $\left(s \neq k\right)$. Consequently X_k is an eigenvector and λ_k is an eigenvalue of matrix .

Example. Let matrix of second order have eigenvectors

$$X_1 = \begin{pmatrix} 1 \\ 2 \end{pmatrix}, \quad X_2 = \begin{pmatrix} 1 \\ 3 \end{pmatrix},$$

which correspond to eigenvalues $\lambda_1 = 1$, $\lambda_2 = 2$.

We find projectors of matrix A $P_1 = \begin{pmatrix} 3 & -1 \\ 6 & -2 \end{pmatrix}$,

$$P_2 = \begin{pmatrix} -2 & 1 \\ -6 & 3 \end{pmatrix}$$

And the matrix A:

$$A = \lambda_1 P_1 + \lambda_2 P_2 = 1 \cdot \begin{pmatrix} 3 & -1 \\ 6 & -2 \end{pmatrix} + 2 \cdot \begin{pmatrix} -2 & 1 \\ -6 & 3 \end{pmatrix} = \begin{pmatrix} -1 & 1 \\ -6 & 4 \end{pmatrix}.$$

5. Polynomial of a matrix

Let A be a square matrix of $m \times m$ size. We shall take a random polynomial

$$\varphi(z) = a_0 + a_1 z + a_2 z^2 + \ldots + a_k z^k.$$

Definition. Polynomial $\varphi(A)$ of matrix   is matrix

$$\varphi(A) = a_0 E + a_1 A + a_2 A^2 + \ldots + a_k A^k. \qquad (40)$$

Let X_s be an eigenvector of matrix   , which corresponds to eigenvalue λ_s. Then the following equalities are true

$$E X s = X_s, \quad A X_s = \lambda_s X_s, \quad A^2 X_s = \lambda_s^2 X_s, \ldots,$$
$$A^k X_s = \lambda_s^k X_s.$$

Here we find equality

$$\varphi(A) X_s = \left(a_0 E + a_1 A + a_2 A^2 + \ldots + a_k A^k \right) X_s =$$
$$= a_0 X_s + a_1 \lambda_s X_s + a_2 \lambda_s^2 X_s + \ldots + a_k \lambda_s^k X_s =$$
$$= \left(a_0 + a_1 \lambda_s + a_2 \lambda_s^2 + \ldots + a_k \lambda_s^k \right) X_s = \varphi(\lambda_s) X_s.$$

From here the following theorem comes.

Theorem. Polynomial   is a matrix which has the same vectors   as matrix   corresponding to eigenvalues $\varphi(\lambda_s)$.

Let $f(\lambda)$ be a characteristic polynomial of matrix

$$f(\lambda) \equiv \det(E \lambda - A).$$

It is followed by Hamilton-Cayley theorem.

Theorem. Any square matrix is a root of its characteristic polynomial [2].

Proof. Let matrix A have different eigenvalues $\lambda_1, \lambda_2, ..., \lambda_m$, which are corresponded by linearly independent eigenvectors $X_1, X_2, ..., X_m$. Eigenvalues are roots of a characteristic polynomial, i.e. $f(\lambda_s) = 0$ $(s = 1, 2, ..., m)$. We have equalities

$$f(A)X_s = f(\lambda_s)X_s = 0 \quad (s = 1, 2, ..., m).$$

(41)

We shall introduce matrix T, columns of which are eigenvectors X of matrix

$$T = (X_1 X_2 ... X_m), \quad \det T \neq 0.$$

Out of equality (41) we receive matrix equality

$$f(A)T = 0.$$ (42)

Since matrix   is non-degenerate then there is inverse matrix T^{-1}. We shall multiply equality (42) by matrix and come to equality $f(A) = 0$.

Example. We shall check validity of Hamilton-Cayley theorem on the example of a matrix of the second order

$$A = \begin{pmatrix} 1 & 2 \\ 3 & 4 \end{pmatrix}, \quad \det(E\lambda - A) \equiv \begin{vmatrix} \lambda - 1 & -2 \\ -3 & \lambda - 4 \end{vmatrix} = \lambda^2 - 5\lambda - 2.$$

We shall find values of matrix polynomial from matrix

$$f(A) \equiv A^2 - 5A - 2E = \begin{pmatrix} 7 & 10 \\ 15 & 12 \end{pmatrix} - 5\begin{pmatrix} 1 & 2 \\ 3 & 4 \end{pmatrix} - 2\begin{pmatrix} 1 & 0 \\ 0 & 1 \end{pmatrix} = \begin{pmatrix} 0 & 0 \\ 0 & 0 \end{pmatrix}.$$

As a result we receive $f(A) = 0$.

We shall consider analytical function $\varphi(z)$, presented by Taylor series

$$\varphi(z) = a_0 + a_1 z + a_2 z^2 + ... + a_k z^k + ...,$$

(43)

which is converged in circle $|z| < R$, containing all eigenvalues $\lambda_1, \lambda_2, ..., \lambda_m$ of matrix A. Analytical function $\varphi(A)$ from matrix   is found by formula

$$\varphi(A) = a_0 E + a_1 A + a_2 A^2 + ... + a_k A^k + ...$$

(44)

From properties of projectors P_k of matrix   and formula (39) for orders of matrix   comes

$$A^n = \lambda_1^k P_1 + \lambda_2^k P_2 + ... + \lambda_m^k P_m.$$ (45)

Row (44) can be written with the help of formula (45)

$$\phi(A) = \sum_{k=0}^{\infty} a_k \left(\lambda_1^k P_1 + \lambda_2^k P_2 + ... + \lambda_m^k P_m \right) = \sum_{s=1}^{m} \left(\sum_{k=0}^{\infty} a_k \lambda_s^k \right) P_s = \sum_{s=1}^{m} \phi(\lambda_s) P_s.$$

Finally we receive formula for an analytical function of matrix

$$\varphi(A) = \sum_{s=1}^{m} \varphi(\lambda_s) P_s.$$

(46)

Matrix $\varphi(A)$ has the same eigenvectors as matrix A and they coincide with eigenvalues $\varphi(\lambda_1), \varphi(\lambda_2), ..., \varphi(\lambda_m)$. Presentation of a function of matrix (46) is called spectral expansion of the function of matrix with a simple structure, when the number of eigenvectors is equal to the order of matrix [3].

Example. We shall find spectral expansion of the function of matrix

$$A = \begin{pmatrix} 1 & 2 \\ 4 & 3 \end{pmatrix}.$$

We shall find eigenvalues λ_1, λ_2 of matrix from equation

$$\det(E\lambda - A) \equiv \begin{vmatrix} \lambda - 1 & -2 \\ -4 & \lambda - 3 \end{vmatrix} = \lambda^2 - 4\lambda - 5 = 0; \quad \lambda_1 = 5,$$

$$\lambda_2 = -1$$

and corresponding to them eigenvectors

$$X_1 = \begin{pmatrix} 1 \\ 2 \end{pmatrix}, \quad X_2 = \begin{pmatrix} -1 \\ 1 \end{pmatrix}.$$

We are building matrix T, T^{-1} and projectors P_1, P_2

$$T = \begin{pmatrix} 1 & -1 \\ 2 & 1 \end{pmatrix}, \quad T^{-1} = \frac{1}{3} \begin{pmatrix} 1 & 1 \\ -2 & 1 \end{pmatrix}$$

$$P_1 = \begin{pmatrix} 1 \\ 2 \end{pmatrix}\begin{pmatrix} \frac{1}{3} & \frac{1}{3} \end{pmatrix} = \frac{1}{3}\begin{pmatrix} 1 & 1 \\ 2 & 2 \end{pmatrix},$$

$$P_2 = \begin{pmatrix} -1 \\ 1 \end{pmatrix}\begin{pmatrix} \frac{-2}{3} & \frac{1}{3} \end{pmatrix} = \frac{1}{3}\begin{pmatrix} 2 & -1 \\ -2 & 1 \end{pmatrix}.$$

By formula (46) we find general expression for the function of matrix A

$$\varphi(A) = \varphi(5)\frac{1}{3}\begin{pmatrix} 1 & 1 \\ 2 & 2 \end{pmatrix} + \varphi(-1)\frac{1}{3}\begin{pmatrix} 2 & -1 \\ -2 & 1 \end{pmatrix}.$$

In a particular case for function $\varphi(z) = z^n$ we receive equality

$$A^n = 5^n\frac{1}{3}\begin{pmatrix} 1 & 1 \\ 2 & 2 \end{pmatrix} + (-1)^n\frac{1}{3}\begin{pmatrix} 2 & -1 \\ -2 & 1 \end{pmatrix}.$$

For function $\varphi(z) = e^{zt}$ we receive formula

$$e^{At} = e^{5t}\frac{1}{3}\begin{pmatrix} 1 & 1 \\ 2 & 2 \end{pmatrix} + e^{-t}\frac{1}{3}\begin{pmatrix} 2 & -1 \\ -2 & 1 \end{pmatrix}.$$

We shall find inverse matrix A^{-1} with the help of function $\varphi(z) = \frac{1}{z}$

$$A^{-1} = \frac{1}{15}\begin{pmatrix} 1 & 1 \\ 2 & 2 \end{pmatrix} - \frac{1}{3}\begin{pmatrix} 2 & -1 \\ -2 & 1 \end{pmatrix} = \frac{1}{5}\begin{pmatrix} -3 & 2 \\ 4 & -1 \end{pmatrix}.$$

6. Lagrange formula for the function of a matrix

Let matrix A have a simple structure and the number of linearly independent eigenvectors be equal to the matrix order. As follows from formula (46), values of two analytical functions $\varphi(A)$, $\psi(A)$ will coincide if the value of functions $\varphi(z)$, $\psi(z)$ coincides on the spectrum of matrix , i.e.

$$\varphi(\lambda_k) = \psi(\lambda_k) \quad (k = 1,2,...,m).$$

At this $\varphi(A) \equiv \psi(A)$.

At the given function of matrix we shall build interpolation polynomial $g(z)$, which in points λ_k takes on values $\varphi(\lambda_k)$

$$g(z) = \frac{(z-\lambda_2)(z-\lambda_3)...(z-\lambda_m)}{(\lambda_1-\lambda_2)(\lambda_1-\lambda_3)...(\lambda_1-\lambda_m)} \varphi(\lambda_1) +$$

$$+ \frac{(z-\lambda_1)(z-\lambda_3)...(z-\lambda_m)}{(\lambda_2-\lambda_1)(\lambda_2-\lambda_3)...(\lambda_2-\lambda_m)} \varphi(\lambda_2) + ... +$$

$$+ \frac{(z-\lambda_1)(z-\lambda_2)...(z-\lambda_{m-1})}{(\lambda_m-\lambda_1)(\lambda_m-\lambda_2)...(\lambda_m-\lambda_{m-1})} \varphi(\lambda_m).$$

We shall put matrix instead of z and receive formula for function of matrix

$$\varphi(A) = \frac{(A - \lambda_2 E)(A - \lambda_3 E)...(A - \lambda_m E)}{(\lambda_1 - \lambda_2)(\lambda_1 - \lambda_3)...(\lambda_1 - \lambda_m)} \varphi(\lambda_1) +$$

$$+ \frac{(A - \lambda_1 E)(A - \lambda_3 E)...(A - \lambda_m E)}{(\lambda_2 - \lambda_1)(\lambda_2 - \lambda_3)...(\lambda_2 - \lambda_m)} \varphi(\lambda_2) + ... +$$

$$+ \frac{(A - \lambda_1 E)(A - \lambda_2 E)...(A - \lambda_{m-1} E)}{(\lambda_m - \lambda_1)(\lambda_m - \lambda_2)...(\lambda_m - \lambda_{m-1})} \varphi(\lambda_m).$$

(47)

Comparing the received formula with formula (46) we receive expression for projectors P_k $(k = 1,2,...,m)$ via matrix A and eigenvalues λ_k of matrix

$$P_k = \prod_{s=1, s \neq k}^{m} \frac{A - \lambda_s E}{\lambda_k - \lambda_s} .$$

(48)

So all projectors of matrix are polynomials of $(m-1)$-order of matrix .

Theorem. We shall find projectors of matrix

$$A = \begin{pmatrix} 1 & 2 \\ 4 & 3 \end{pmatrix}, \quad \lambda_1 = 5, \quad \lambda_2 = -1.$$

According to formula (48) we will receive expressions

$$P_1 = \frac{A - \lambda_2 E}{\lambda_1 - \lambda_2} = \frac{1}{6}\begin{pmatrix} 2 & 2 \\ 4 & 4 \end{pmatrix},$$

$$P_2 = \frac{A - \lambda_1 E}{\lambda_2 - \lambda_1} = \frac{-1}{6}\begin{pmatrix} -4 & 2 \\ 4 & -2 \end{pmatrix}.$$

Let $f(z)$ be a characteristic polynomial of matrix A

$$f(z) \equiv (z - \lambda_1)(z - \lambda_2)..(z - \lambda_m).$$

From Hamilton-Cayley theorem the following equality follows

$$f(A) \equiv (A - \lambda_1 E)(A - \lambda_2 E)..(A - \lambda_m E).$$

From formula (48) it follows that

$$(A - \lambda_k E)P_k = 0; \quad P_k(A - \lambda_k E) = 0.$$

$$(49)$$

Consequently, all rows of matrix P_k are proportional to the right eigenvector X_k of matrix A, and all rows of projector are proportional to the left eigenvector of matrix .

We should note that all ranks of projectors $(k = 1,2,...,m)$ *are equal to one.*

We shall demonstrate one more way to build projectors of matrix . We shall consider auxiliary function

$$\varphi(z) = \frac{1}{\lambda - z}. \tag{50}$$

At big value $|\lambda|$ function $\varphi(z)$ is analytical in the circle which contains eigenvalues $\lambda_1, \lambda_2, ..., \lambda_m$ of matrix A. By formula (46) we find equality

$$(E\lambda - A)^{-1} = \frac{1}{\lambda - \lambda_1} P_1 + \frac{1}{\lambda - \lambda_2} P_2 + ... + \frac{1}{\lambda - \lambda_m} P_m. \tag{51}$$

For finding projectors we can use residues

$$P_k = res(E\lambda - A)^{-1} \quad \text{при} \quad \lambda = \lambda_k. \tag{52}$$

The left part of equality (51) can be calculated with the help of attaching matrix $B(\lambda)$, elements of which $b_{ks}(\lambda)$ are algebraic additions of element $\delta_{sk}\lambda - a_{sk}$ of matrix $E\lambda - A$ $(\delta_{kk} = 1; \ \delta_{sk} = 0; \ s \neq k)$.

For the inverse matrix we have equality

$$(E\lambda - A)^{-1} = \frac{B(\lambda)}{f(\lambda)}; \quad f(\lambda) = \det(E\lambda - A), \tag{53}$$

where $f(x)$ is a characteristic polynomial of matrix . Finding residues in points , we come to the known formulas for projectors of matrix [2]

$$P_k = \underset{\lambda = \lambda_k}{res} \frac{B(\lambda)}{f(\lambda)}; \quad P_k = \frac{B(\lambda_k)}{f'(\lambda_k)} \quad (k = 1, 2, ..., m);$$

$$f'(\lambda) \equiv \frac{df(\lambda)}{d\lambda}. \tag{54}$$

Example. We shall find projectors of matrix

$$A = \begin{pmatrix} 0 & 2 \\ -1 & 3 \end{pmatrix}.$$

We find characteristic polynomial $f(\lambda)$, eigenvalues λ_1, λ_2 of matrix A and mounting matrix $B(\lambda)$

$$f(\lambda) = \det(E\lambda - A) \equiv \begin{vmatrix} \lambda & -2 \\ 1 & \lambda - 3 \end{vmatrix} \equiv \lambda^2 - 3\lambda + 2;$$

$$f(\lambda) = 0; \; \lambda_1 = 1, \; \lambda_2 = 2; \quad B(\lambda) = \begin{pmatrix} \lambda - 3 & 2 \\ -1 & \lambda \end{pmatrix}.$$

By formula (54) we calculate projectors

$$P_1 = \frac{B(1)}{f'(1)} = \begin{pmatrix} 2 & -2 \\ 1 & -1 \end{pmatrix}; \quad P_2 = \frac{B(2)}{f'(2)} = \begin{pmatrix} -1 & 2 \\ -1 & 2 \end{pmatrix}.$$

Any analytical function $\varphi(A)$ of matrix   is found by formula (46)

.

In particular at $\varphi(z) = \ln z$ we receive

$$\ln A = \ln 2 \cdot \begin{pmatrix} -1 & 2 \\ -1 & 2 \end{pmatrix}.$$

7. Cauchy formula for the formula of matrix

Let analytical function $\varphi(z)$ be analytical in some simply connected region $\mathcal{Д}$, containing spectrum of matrix A, a set of all eigenvalues of matrix . We shall find the value of analytical function by formula (46), where is the value of function , analytically extended from some point of region . Let closed without intersections contour Γ span the whole spectrum of matrix . In this case we can find the values of function $\varphi(\lambda)$ with the help of Cauchy formula

$$\varphi(\lambda) = \frac{1}{2\pi i} \int_\Gamma \varphi(z)(z - \lambda)^{-1} dz. \tag{55}$$

Formula (46) can be written like

$$\varphi(A) = \sum_{k=1}^{m} \frac{1}{2\pi i} \int_\Gamma \frac{\varphi(z)}{z - \lambda_k} dz \cdot P_k,$$

which, while concerning equality (51), looks like

$$\varphi(A) = \frac{1}{2\pi i} \int_\Gamma \varphi(z)(Ez - A)^{-1} dz. \tag{56}$$

Formula (56) is analogues to Cauchy formula (55). Instead of value λ we put matrix . This formula is true for analytical function $\varphi(A)$ of matrix . Due to formula (56) we can see that the analytical function of the matrix is a continuous function of matrix , i.e.

during continuous changing of the elements of matrix A the elements of matrix $\varphi(A)$ are changing continuously as well.

Out of formula (51) it follows that projectors of matrix can be found by formula

$$P_k = \frac{1}{2\pi i} \int_{\Gamma_k} (Ez - A)^{-1} dz \quad (k = 1, 2, ..., m),$$

$$(57)$$

where Γ_k is a closed contour spanning only point $z = \lambda_k$ and not containing any other points of spectrum $z = \lambda_s$ $(s \neq k; \ s = 1, 2, ..., m)$ inside.

If some contour γ spans together several points of spectrum $\lambda_1, ..., \lambda_s$, we find the sum of the corresponding projectors by formula

$$P_1 + ... + P_s = \frac{1}{2\pi i} \int_{\gamma} (Ez - A)^{-1} dz.$$

While introducing formula (56) it is assumed that matrix has simple eigenvalues. We shall give another derivation of formula (56), which does not use assumptions about simplicity of the structure of matrix .

Let function $\varphi(z)$ be analytical in closed circle $|z| \leq R$ и $\|A\| < R$. At that all eigenvalues of matrix are located inside circle $|z| \leq R$. We shall expand function $\varphi(z)$ by Taylor series

$$\varphi(z) = \varphi(0) + \frac{\varphi'(0)}{1!} z + \frac{\varphi''(0)}{2!} z^2 + \ldots + \frac{\varphi^{(n)}(0)}{n!} z^n + \ldots,$$

where coefficients are found by Cauchy formula

$$\frac{\varphi^{(n)}(0)}{n!} = \frac{1}{2\pi i} \int\limits_{|z|=R} \frac{\varphi(z)}{z^{n+1}} dz \quad (n = 0,1,2,\ldots).$$

(58)

Expansion of function $\varphi(A)$ of matrix A into series

$$\varphi(A) = \varphi(0)E + \frac{\varphi'(0)}{1!} A + \frac{\varphi''(0)}{2!} A^2 + \ldots + \frac{\varphi^{(n)}(0)}{n!} A^n + \ldots$$

can be written like

$$\varphi(A) = \frac{1}{2\pi i} \int\limits_{|z|=R} \varphi(z) \left(\frac{E}{z} + \frac{A}{z^2} + \frac{A^2}{z^3} + \ldots + \frac{A^n}{z^{n+1}} + \ldots \right) dz.$$

(59)

Following the assumption about the norm of matrix   , series

$$(Ez - A)^{-1} = \frac{E}{z} + \frac{A}{z^2} + \frac{A^2}{z^3} + \ldots + \frac{A^n}{z^{n+1}} + \ldots$$

is uniformly reduced on circle $|z| = R$ and can be termwise. We receive the formula for the random analytical function of matrix

$$\varphi(A) = \frac{1}{2\pi i} \int_{|z|=R} f(z)(Ez - A)^{-1} dz. \tag{60}$$

This formula is a matrix analogue of Cauchy formula (55). Contour $|z| = R$ can be replaced by another contour Γ, which contains the spectrum of matrix A. The elements of matrix $\varphi(A)$ are analytically dependent on the elements of matrix .

Let function $\varphi(z)$ be analytically prolonged from circle $|z| \leq R$ into some simply connected region $Д$ with limit , which contains all eigenvalues $\lambda_1, \lambda_2, ..., \lambda_m$ of matrix . At that we come to formula (55), where the assumption about simplicity of the structure of matrix is not used. With the help of formula (55) we can receive spectral expansion of the function of the matrix in the general case.

Let matrix have eigenvalues $\lambda_1, ..., \lambda_q$ of $m_1, .., m_q$-multiple, $m_1 + ... + m_q = m$. Characteristic polynomial $f(z)$ has expansion into factors

$$f(z) \equiv \det(Ez - A) = (z - z_1)^{m_1} ... (z - z_q)^{m_q}.$$

We find analytical representation of inverse matrix $(Ez - A)^{-1}$ with the help of expansion of the matrix elements into the simplest fractions

$$\left(\mathrm{E}z-\mathrm{A}\right)^{-1}=\frac{B(z)}{f(z)}=\sum_{k=1}^{q}\left(\frac{B_{k1}}{z-\lambda_k}+\frac{B_{k2}\cdot 1!}{\left(z-\lambda_k\right)^2}+...+\frac{B_{km_k}\left(m_k-1\right)!}{\left(z-\lambda_k\right)^{m_k}}\right).$$

By Cauchy formula we find the main formula for the function of
random matrix [2]

$$\phi(\mathrm{A})=\sum_{k=1}^{q}\left(f(\lambda_k)B_{k1}+f'(\lambda_k)B_{k2}+...+f^{(m-1)}(\lambda_k)B_{km_k}\right),$$

$$(61)$$

which can be called spectral expansion of the function of matrix A. Matrixes B_{ks} are called components of matrix . They do not depend on the view of matrix $\varphi(z)$ and can be found completely by matrix . Matrixes can be expressed by adjoined matrix $B(z)$

$$B_{ks}=\lim_{z\to\lambda_k}\frac{1}{r!(s-1)!}\frac{d^r}{dz^r}\left(\frac{B(z)(z-\lambda_k)^{m_k}}{f(z)}\right)$$

$$\left(k=1,...,q;\ s=1,...,m_k;\ r\equiv m_k-s\right).$$

Matrixes can be found by interpolation formulas when the value of interpolation polynomial $g(z)$ coincides on the spectrum of matrix (concerning the multiplicity of eigenvalues) with the values of function , i.e. equalities are fulfilled

$$g(\lambda_k)=\varphi(\lambda_k),\ \ g'(\lambda_k)=\varphi'(\lambda_k),\ ...,$$

$$g^{(m_k-1)}(\lambda_k)=\phi^{(m_k-1)}(\lambda_k)\ \ \ \left(k=1,...,q\right).$$

In particular cases we can find such simple functions $\varphi(z)$, polynomials as a rule, when we can easily calculate the left part in formula (61), and then sequentially find components of matrix A. We shall note that some matrixes B_{ks} can appear to be zero.

Example. We shall apply formula (61) to find the function of matrix

$$A = \begin{pmatrix} -1 & 2 \\ -2 & 3 \end{pmatrix}$$

with multiple eigenvalue $\lambda_1 = 1$, $m_1 = 2$. For analytical function which is analytical in point $z = 1$ we have expression (61)

$$\varphi(A) = \varphi(1)B_1 + \varphi'(1)B_2, \tag{62}$$

where matrixes B_1, B_2 do not depend on function .

We shall take two functions $\varphi_1(z) \equiv 1$, $\varphi_2(z) \equiv z$. From formula (62) we receive equalities

$$\begin{cases} E = 1 \cdot B_1 + 0 \cdot B_2 \\ A = 1 \cdot B_1 + 1 \cdot B_2 \end{cases}$$

from which we find: $B_1 = E$, $B_2 = A - E$. Finally we find the formula for the function of matrix

$$\varphi(A) = \varphi(1) \cdot E + \varphi'(1)(A - E)$$

or

$$\varphi(A) = \varphi(1)\begin{pmatrix} 1 & 0 \\ 0 & 1 \end{pmatrix} + \varphi'(1)\begin{pmatrix} -2 & 2 \\ -2 & 2 \end{pmatrix}.$$

In order to check it we shall calculate matrixes A^2, A^{-1} directly and by formula (62). We have

$$A^2 = \begin{pmatrix} -1 & 2 \\ -2 & 3 \end{pmatrix}\begin{pmatrix} -1 & 2 \\ -2 & 3 \end{pmatrix} = \begin{pmatrix} -3 & 4 \\ -4 & 5 \end{pmatrix}, \quad A^{-1} = \begin{pmatrix} 3 & -2 \\ 2 & -1 \end{pmatrix}.$$

By formula (62) we find the same values

$$A^2 = 1 \cdot \begin{pmatrix} 1 & 0 \\ 0 & 1 \end{pmatrix} + 2 \cdot \begin{pmatrix} -2 & 2 \\ -2 & 2 \end{pmatrix} = \begin{pmatrix} -3 & 4 \\ -4 & 5 \end{pmatrix};$$

$$A^{-1} = 1 \cdot \begin{pmatrix} 1 & 0 \\ 0 & 1 \end{pmatrix} - 1 \cdot \begin{pmatrix} -2 & 2 \\ -2 & 2 \end{pmatrix} = \begin{pmatrix} 3 & -2 \\ 2 & -1 \end{pmatrix}.$$

Remark. If we take function $\varphi(z) \equiv z$ in formula (61), we shall receive the well-known representation of matrix A via its components

$$A = \sum_{k=1}^{q} (\lambda_k B_{k1} + B_{k2}). \tag{63}$$

Example. We shall consider matrix of the third order

$$A = \begin{pmatrix} -2 & -3 & 3 \\ -4 & -3 & 4 \\ -6 & -6 & 7 \end{pmatrix}.$$

Characteristic polynomial

$$f(z) = \det(Ez - A) = z^3 - 2z + z = (z-1)^2 z$$

has double multiplex root $z_1 = 1$ and single multiplex root $z_2 = 0$.

Any analytical function $\varphi(\mathrm{A})$ of the matrix looks like

$$\varphi(\mathrm{A}) = \varphi(1)B_{11} + \varphi'(1)B_{12} + \varphi(0)B_{21}.$$

We shall choose polynomial functions

$$\varphi_1(z) = 2z - z^2; \quad \varphi_2(z) = z^2 - z; \quad \varphi_3(z) = (z-1)^2.$$

At that we receive components of matrix A

$$B_{11} = 2\mathrm{A} - \mathrm{A}^2 = \begin{pmatrix} -2 & -3 & 3 \\ -4 & -3 & 4 \\ -6 & -6 & 7 \end{pmatrix} = \mathrm{A};$$

$$B_{12} = \mathrm{A}^2 - \mathrm{A} = \begin{pmatrix} 0 & 0 & 0 \\ 0 & 0 & 0 \\ 0 & 0 & 0 \end{pmatrix};$$

$$B_{21} = (\mathrm{A} - \mathrm{E})^2 = \begin{pmatrix} 3 & 3 & -3 \\ 4 & 4 & -4 \\ 6 & 6 & -6 \end{pmatrix}.$$

In this example $H_{12} = 0$, i.e. matrix   has a simple structure. Any analytical function of matrix   looks like

$$\varphi(A) = \varphi(1) \cdot \begin{pmatrix} -2 & -3 & 3 \\ -4 & -3 & 4 \\ -6 & -6 & 7 \end{pmatrix} + \varphi(0) \begin{pmatrix} 3 & 3 & -3 \\ 4 & 4 & -4 \\ 6 & 6 & -6 \end{pmatrix}.$$

At $\varphi(z) = z^n$ $(n = 1,2,3,...)$ we receive

$$A^n = A \quad (n = 1,2,...)$$

i.e. matrix A is a projector.

Remark. Let matrix   have multiple eigenvalues. Then analytical function $\varphi(A)$ of matrix is found by formula (61). At that the components of matrix B_{ks} are found by formula

$$B_{k1} = \frac{1}{2\pi i} \int_{\Gamma_k} (Ez - A)^{-1} dz \quad (k = 1,...,q)$$

and are, consequently, projectors of matrix   .

$$B_{k1} = P_k \qquad ,$$

contour Γ_k spans only point $z = \lambda_k$ on complex plane z. The rest components of matrix   can be found by formula

$$B_{ks} = \frac{1}{(s-1)!} \cdot \frac{1}{2\pi i} \int_{\Gamma_k} (Ez - A)^{-1} (z - \lambda_k)^s dz$$

$$(k = 1,...,q; \ s = 1,...,m_k). \quad (64)$$

All components   of matrix   are analytical functions of matrix   and thus they are all adjusting

$$B_{ks} \cdot B_{ij} = B_{ij} \cdot B_{ks} \quad (k,i = 1,...,q; \ s,j = 1,...,m_k).$$

8. Subspaces

To understand the following results we need to introduce the notion of subspace.

Let L be some set of vectors in m-dimensional space. If any linear combination of vectors from   is also a vector from   , then the set of vectors   is called subspace if the following conditions are fulfilled:

1. If $X_1 \in L$, $X_2 \in L$, then $X_1 + X_2 \in L$.

2. If $X \in L$, λ is a random value, then $\lambda X \in L$.

Zero vector 0 belongs to any subspace.

We can give another definition of subspace   .

We have some linearly independent vectors $X_1, X_2, ..., X_q$. The set of all vectors like

$$X = \alpha_1 X_1 + \alpha_2 X_2 + ... + \alpha_q X_q$$

for random values of coefficients $\alpha_1, \alpha_2, ..., \alpha_q$ creates subspace   of q dimention. If in subspace   has   of linearly independent vectors, and any $q+1$ vectors from   are linearly dependent, then the integer   is called dimension of the subspace and defined as $\dim L$. Subspace   , found by the linear combination of vectors   , is also called subspace which is stretched on vectors   or a linear shell of vectors

.

We shall take random vector X_0. To find out if vector   belongs to subspace L, we should find the solution of equation system

$$\alpha_1 X_1 + \alpha_2 X_2 + ... + \alpha_q X_q = X_0 \qquad (65)$$

relatively unknown $\alpha_1, \alpha_2, ..., \alpha_q$. If equation system (65) has a solution, then $X_0 \in L$. If equation system (65) does not have a solution, then $X_0 \overline{\in} L$. Linearly called vectors $X_1, X_2, ..., X_q$ are called the basis of subspace calls   , stretched on vectors   . In formula (65) coefficients   are   called coordinates of vector   .

It is easy to prove that the sum of $L_1 \cup L_2$ of two subspaces L_1, L_2 and intersection $L_1 \cap L_2$ of these subspaces are also subspaces. At that the equation [2] is true

$$\dim L_1 + \dim L_2 = \dim L_1 \cup L_2 + \dim L_1 \cap L_2.$$

$$(66)$$

if   ,   are two subspaces in m-dimensional vector subspace   and there is no eigenvector, besides a zero vector, which belongs to every subspace   ,   , then subspace $L = L_1 \cup L_2$ is called the direct sum of subspaces   ,   and defined as

$$L = L_1 + L_2.$$

At that any vector $X \in L$ is only represented as the sum of two vectors X_1, X_2:

$$X = X_1 + X_2, \quad X_1 \in L_1, \quad X_2 \in L_2. \tag{67}$$

The basis of subspace L can be received due to the union of bases of subspaces L_1, $Д_2$.

For setting specified subspaces we can use notions of a nullified subspace and a range of values of the matrix.

Let matrix T of $m \times m$-size have a rank equal to имеет q $(q \le m)$. The set of vectors X, satisfying equation

$$TX = 0 \tag{68}$$

create a subspace, которое which is called the nucleus of matrix , or a nullified subspace and defined like KerT. It is obvious that the following equality is performed

$$\dim Ker T = m - rang T = m - q.$$

$$\tag{69}$$

From matrix   we can choose   of linearly independent lines and unite them in matrix T_1 of $q \times m$-size and write equation (68) like

$$T_1 X = 0.$$

The set of vectors   , represented like

$$X = TY, \quad rangT = q, \tag{70}$$

where Y is some vector of m-dimension, is called the image of matrix T, of the range of values and defined as $I_m T$. The image of matrix   is a subspace the basis of which is formed by any q linearly independent vectors-columns of matrix   . The obvious equality is true

$$\dim I_m T = rangT. \tag{71}$$

So for any square matrix   of   -order the following formula is true

$$\dim I_m T + \dim KerT = m. \tag{72}$$

In the case when $rangT = m$, we shall receive

$$\dim I_m T = m, \quad \dim KerT = 0.$$

For $0 < q < m$ we can choose можно   linearly independent rows of matrix   , create from them matrix T_2 of $m \times q$ size and set vector и X (70) by equality

$$X = T_2 Z, \quad \dim Z = q. \tag{73}$$

For   we can set a specified subspace by equation system (68) or (70). We can always transform equations (68) to equation (70) and vice versa.

We can show a simple way of the transformations.

Let subspace L be set by equation system (68). We shall choose in matrix T q linearly independent lines and form from them first   lines of the auxiliary matrix T_3 of $m \times m$ size, and the rest $m-q$ lines will be chosen randomly so that $\det T_3 \neq 0$.

We shall find inverse matrix T_3^{-1}. Since $T_3 T_3^{-1} = E$, as matrix   we can take the matrix which contains the last   rows of matrix   . At that subspace L_2 can be shown like equation system (73).

In fact due to creating $T_1 T_2 = 0$.

Example. Let us set matrix

$$T = \begin{pmatrix} 1 & -1 & 2 \\ 2 & -2 & 4 \\ -1 & 1 & -2 \end{pmatrix}, \quad rang T = 1.$$

We shall create matrix   and find

$$T_3 = \begin{pmatrix} 1 & -1 & 2 \\ 0 & 1 & 0 \\ 0 & 0 & 1 \end{pmatrix}, \quad T_3^{-1} = \begin{pmatrix} 1 & 1 & -2 \\ 0 & 1 & 0 \\ 0 & 0 & 1 \end{pmatrix}.$$

From the last two rows of matrix   we shall create matrix T_2. Subspace $TX = 0$ can be written like

$$X = \begin{pmatrix} 1 & -2 \\ 1 & 0 \\ 0 & 1 \end{pmatrix} Z, \quad \dim Z = 2.$$

Inversely, let subspace L be set by equation system (70). We shall point out q linearly independent rows of matrix T and create from them the fir st   rows of matrix T_4 of $m \times m$ size. The last $m-q$ rows will be chosen randomly so that $\det T_4 \neq 0$. Since $T_4^{-1} T_4 = E$, then as matrix T_1 we can take the low   lines of matrix T_4^{-1} and write subspace like equations the system of equations (68).

   Example. Let subspace   be set by equation system $X = TY$, where matrix   is written like formula

$$T = \begin{pmatrix} 1 & -1 & 2 \\ 2 & -2 & 4 \\ -1 & 1 & -2 \end{pmatrix}, \quad rangT = 1.$$

We shall create matrix   and find matrix

$$T_4 = \begin{pmatrix} 1 & 0 & 0 \\ 2 & 1 & 0 \\ -1 & 0 & 1 \end{pmatrix}, \quad T_4^{-1} = \begin{pmatrix} 1 & 0 & 0 \\ -2 & 1 & 0 \\ 1 & 0 & 1 \end{pmatrix}.$$

As matrix B T_1 the low two lines of matrix T_4^{-1} can be taken and subspace L can be written like the system of equations

$$\begin{pmatrix} -2 & 1 & 0 \\ 1 & 0 & 1 \end{pmatrix} X = 0.$$

9. Invariant subspaces of the matrix similar to the block matrix

Let matrix A be similar to block diagonal matrix Λ, i.e. there is such matrix T that

$$T^{-1}AT = \Lambda, \quad \Lambda = \begin{pmatrix} \Lambda_1 & 0 & \dots & 0 \\ 0 & \Lambda_2 & \dots & 0 \\ \dots & \dots & \dots & \dots \\ 0 & 0 & \dots & \Lambda_q \end{pmatrix}.$$

(74)

Matrixes Λ_k are square matrixes of – квадратные матрицы m_k $\left(m_1 + m_2 + \dots + m_q\right) = m$ -order. We shall divide T, T^{-1} into blocks of $m \times m_k$, $m_k \times m$ orders correspondingly

$$T = \begin{pmatrix} T_1 & T_2 & \dots & T_q \end{pmatrix}, \quad T^{-1} = \begin{pmatrix} u_1 \\ u_2 \\ \dots \\ u_q \end{pmatrix}.$$

Let function $\varphi(z)$ be analytical in some area D, containing the spectrum of matrix   . We shall assume that spectrums of matrixes   are separated and can be included into simple non-intersected contour Γ_k. Let contour Γ span the spectrum of matrix   . We shall find function $\varphi(A)$ by formula

$$(75)$$

Since for the block matrix the following equality is true

$$
(Ez - \Lambda)^{-1} =
\begin{pmatrix}
(Ez - \Lambda_1)^{-1} & 0 & \ldots & 0 \\
0 & (Ez - \Lambda_2)^{-1} & \ldots & 0 \\
\ldots & \ldots & \ldots & \ldots \\
0 & 0 & \ldots & (Ez - \Lambda_q)^{-1}
\end{pmatrix}
,
$$

we shall receive for the function of matrix A

$$
\varphi(A) = T
\begin{pmatrix}
f(\Lambda_1) & 0 & \ldots & 0 \\
0 & f(\Lambda_2) & \ldots & 0 \\
\ldots & \ldots & \ldots & \ldots \\
0 & 0 & \ldots & f(\Lambda_q)
\end{pmatrix}
T^{-1}
$$

the well-known formula for the function of block matrix

$$\varphi(A) = T_1 \varphi(\Lambda_1) u_1 + T_2 \varphi(\Lambda_2) u_2 + \ldots + T_q \varphi(\Lambda_q) u_q.$$

$$(76)$$

Example. Let matrix   be diagonal

$$
A =
\begin{pmatrix}
\alpha_1 & 0 & \ldots & 0 \\
0 & \alpha_2 & \ldots & 0 \\
\ldots & \ldots & \ldots & \ldots \\
0 & 0 & \ldots & \alpha_m
\end{pmatrix}
.
$$

Then we have equation

$$\varphi(A) = \begin{pmatrix} \varphi(\alpha_1) & 0 & \dots & 0 \\ 0 & \varphi(\alpha_2) & \dots & 0 \\ \dots & \dots & \dots & \dots \\ 0 & 0 & \dots & \varphi(\alpha_m) \end{pmatrix}.$$

In a particular case we shall receive equality

$$e^{At} = \begin{pmatrix} e^{\alpha_1 t} & 0 & \dots & 0 \\ 0 & e^{\alpha_2 t} & \dots & 0 \\ \dots & \dots & \dots & \dots \\ 0 & 0 & \dots & e^{\alpha_m t} \end{pmatrix}. \tag{77}$$

If matrix A is similar to diagonal matrix

$$T^{-1}AT = \begin{pmatrix} \alpha_1 & 0 & \dots & 0 \\ 0 & \alpha_2 & \dots & 0 \\ \dots & \dots & \dots & \dots \\ 0 & 0 & \dots & \alpha_m \end{pmatrix},$$

then for the demonstrative matrix we receive the following expression

$$e^{At} = T \begin{pmatrix} e^{\alpha_1 t} & 0 & \dots & 0 \\ 0 & e^{\alpha_2 t} & \dots & 0 \\ \dots & \dots & \dots & \dots \\ 0 & 0 & \dots & e^{\alpha_m t} \end{pmatrix} T^{-1}. \tag{78}$$

Any element of matrix e^{At} is a linear combination of demonstrative functions

$$e^{At} = \sum_{k=1}^{m} T_k e^{\alpha_k t} U_k, \qquad (79)$$

where T_k, u_k are constant matrixes.

For the block matrix we shall find expressions for projectors of matrix A by formulas

$$P_k = \frac{1}{2\pi i} \int_{\Gamma_k} \left(Ez - A \right)^{-1} dz = \frac{1}{2\pi i} \int_{\Gamma_k} \sum_{s=1}^{m} T_s \left(Ez - \Lambda_s \right)^{-1} U_s dz =$$

$$= T_k \frac{1}{2\pi i} \int_{\Gamma_k} \left(Ez - \Lambda_k \right)^{-1} dz \, U_k = T_k E_{m_k} U_k = T_k U_k \qquad \left(k = 1, 2, ..., q \right).$$

$$(80)$$

Here E_{m_k} is a unit matrix of m_k-order.

Projectors P_k $\left(k = 1, 2, ..., q \right)$ define linear subspaces $L_k = I_m P_k$. At that the nucleus of matrix will be the subspace of column vectors, orthogonal to the row of matrix U_k^*, i.e. the image of matrix $E - P_k$.

Similarly linear subspaces L_k^* of the space of row vectors L^* are created by linear combinations of the lines of matrix U_k .

The rank of projector is equal to order m_k of matrix Λ_k. as follows out of formula

$$P_k = T_k U_k \qquad (81)$$

the rows of projector P_k are linear combinations of the columns of matrix T_k, and the lines of projector   are linear combinations of the lines of matrix U_k.

If we know projector   , then the basis of the subspace can be found by allocating linearly independent rows of matrix   . Similarly the basis of subspace L_k^* can be found allocating linearly independent lines of matrix   . Following condition

$$U_k T_k = E_{m_k} \qquad (k = 1,2,...,q)$$

we find expansion of projector   into factors (81).

We shall show that subspaces L_k , defined by projectors $(k = 1,2,..,q)$ are invariant subspaces of matrix A by formula (75) for $\varphi(z) \equiv z$:

$$A = \sum_{k=1}^{q} \frac{1}{2\pi i} \int_{\overset{\circ}{A}_k} z (Ez - A)^{-1} dz = \sum_{k=1}^{q} \frac{1}{2\pi i} \int_{\overset{\circ}{A}_k} T_k z (Ez - \Lambda_k)^{-1} U_k dz =$$

$$= \sum_{k=1}^{q} \frac{1}{2\pi i} \int_{\overset{\circ}{A}_k} T_k \Lambda_k (Ez - \Lambda_k)^{-1} U_k dz = \sum_{k=1}^{q} T_k \Lambda_k U_k$$

.

$$(82)$$

As follows from obvious formulas

$$U_k T_k = E_{m_k}, \qquad U_k T_s = 0 \qquad (k \neq s; \ k,s = 1,2,...,q),$$

$$(83)$$

coming out of equations $T^{-1}T = E$, for $X \in L_k$ the following relationship is fulfilled

$$AX = \sum_{k=1}^{q} T_k \Lambda_k U_k X = T_k (\Lambda_k U_k), \quad AX \in L_k.$$

Consequently subspaces L_k are images of projector P_k and are the invariant subspace of matrix A.

As follows from formulas (82), (83) we have equations

$$AT_k = T_k \Lambda_k, \quad U_k A = \Lambda_k U_k \quad (k = 1,2,...,q),$$

$$(84)$$

which generalize notions of the eigenvector and eigenvalue of the matrix. Matrixes T_k, U_k are called eigenmatrixes and matrixes Λ_k are called matrix factors.

10. Invariant subspaces of the matrix with multiple eigenvalues spaces

We shall consider the case when in matrixes Λ_k all eigenvalues are identical and equal to λ_k.

Let matrixes Λ_k be Jordan cells:

$$\Lambda_k = \begin{pmatrix} \lambda_k & 1 & 0 & \dots & 0 \\ 0 & \lambda_k & 1 & \dots & 0 \\ 0 & 0 & \lambda_k & \dots & 0 \\ \dots & \dots & \dots & \dots & \dots \\ 0 & 0 & 0 & \dots & \lambda_k \end{pmatrix} \quad (85)$$

of $m_k \times m_k$ size. Different matrixes Λ_k can have identical eigenvalues. We shall define the columns of matrix T_k as $X_{k1},...,X_{km_k}$. As follows from the first matrix equality (84) the following equalities are true

$$AX_{k1} = \lambda_k X_{k1}, \quad AX_{k2} = \lambda_k X_{k2} + X_{k1},$$
$$AX_{km_k} = \lambda_k X_{km_k} + X_{km_k-1}. \quad (86)$$

Vector X_{k1} is called an eigenvector and vectors $X_{k2},...,X_{km_k}$ are called the attached vectors to   .

Vectors $X_{k1}, X_{k2},...,X_{km_k}$, which are creating invariant subspace L_k of matrix A, correspond to Jordan cells Λ_k. As follow from equality (86) we receive the following formulas

$$(A - \lambda_k E) X_{k1} = 0, \quad (A - \lambda_k E)^2 X_{k2} = 0, \quad ...,$$

$$(A - \lambda_k E)^{m_k} X_{km_k} = 0. \quad (87)$$

Consequently, vectors $X_{k1}, X_{k2}, ..., X_{km_k}$ enter the nucleus of matrix $(A - \lambda_k)^{m_k}$. Thus the following equality is true

$$(A - \lambda_k E)^{m_k} T_k = 0. \quad (88)$$

If a number of Jordan cells have identical eigenvalues then all the vectors of the basis of the invariant subspaces satisfy vector equation

$$(A - \lambda_k E)^{v_k} X = 0, \quad (89)$$

where v_k is the biggest from the order of Jordan cells which have eigenvalue λ_k.

Similarly, the line vectors which are creating the basis of the invariant subspace in the conjugate subspace L^* satisfy equation

$$Y(A - \lambda_k E)^{v_k} = 0. \quad (90)$$

We shall consider in detail the structure of the analytical function of the matrix in the case of multiple roots of the characteristic equation. We assume that in equation (74) all matrixes Λ_n are Jordan cells (85). We shall suppose

$$\Lambda_k = E_{m_k} \lambda_k + F_{m_k}, \quad (91)$$

where matrix F_{m_k} has order m_k. All elements of matrix F_{m_k} are equal to zero excluding unit elements which occupy one diagonal higher than the main one:

$$F_{m_k} = \begin{pmatrix} 0 & 1 & 0 & ... & 0 \\ 0 & 0 & 1 & ... & 0 \\ ... & ... & ... & ... & ... \\ 0 & 0 & 0 & ... & 1 \\ 0 & 0 & 0 & ... & 0 \end{pmatrix}. \qquad (92)$$

By formula (61) we receive an analytical expression for the function of the matrix

$$\varphi(A) = \sum_{k=1}^{q} T \frac{1}{2\pi i} \int_{\Gamma_k} \varphi(z)(E_z - \Lambda)^{-1} dz T^{-1} =$$

$$= \sum_{k=1}^{q} T_k \frac{1}{2\pi i} \int_{\Gamma_k} \varphi(z)(E_{m_k} z - \lambda_k E_{m_k} - F_{m_k})^{-1} dz U_k.$$

$$(93)$$

As the expansion is true

$$\left(E_{m_k}(z - \lambda_k) - F_{m_k}\right)^{-1} = \frac{1}{z - \lambda_k} E_{m_k} + \frac{1}{(z - \lambda_k)^2} F_{m_k} +$$

$$+ \frac{1}{(z - \lambda_k)^3} F_{m_k}^2 + ... + \frac{1}{(z - \lambda_k)^{m_k - 1}} F_{m_k}^{m_k - 1},$$

$$(94)$$

then as follows from formula (93) we receive an analytical image of the function of matrix A

$$\varphi(A) = \sum_{k=1}^{q} T_k (\varphi(\lambda_k) E_{m_k} + \frac{\varphi'(\lambda_k)}{1!} F_{m_k} + \frac{\varphi''(\lambda_k)}{2!} F_{m_k}^2 + . \quad (95)$$

$$+ \ldots + \frac{\varphi^{(m_k-1)}(\lambda_k)}{(m_k-1)!} F_{m_k}^{m_k-1}) U_k$$

Comparing to formula (61) allows to receive an explicit expression for components B_{ks} of matrix

$$B_{k1} = T_k U_k, \quad B_{ks} = T_k F_{m_k}^{s-1} U_k \quad (s = 2,\ldots,m_k).$$
$$(96)$$

These formulas show that

$$I_m B_{ks} \subset I_m B_{k1} \qquad . \qquad (97)$$

If we suppose by definition that $F_{m_k}^0 = E_{m_k}$, then formulas (96) can be united in one formula

$$B_{ks} = T_k F_{m_k}^{s-1} U_k \quad (s = 1,2,\ldots,m_k). \qquad (98)$$

Since $U_k T_k = E_{m_k}$, $U_k T_s = 0$ $(k \neq s)$, then we receive equalities for the components of matrix

$$B_{ks} B_{nj} = T_k F_{m_k}^{s-1} U_k T_n F_{m_k}^{j-1} U_n = \delta_{kn} T_k F_{m_k}^{s+j-2} U_k = \delta_{kn} B_{k,s+j-1}.$$

It should be considered that $F_{m_k}^{m_k} = 0$, i.e. $B_{ks} = 0$ $(s > m_k)$.

Many of the function properties are shown in works [1, 4].

11. An algebraic way of building projectors

We shall show one way of building projectors of the matrix in the case when a minimal polynomial is known

$$\psi(z) = (z - \lambda_1)^{v_1} (z - \lambda_2)^{v_2} ... (z - \lambda_q)^{v_q}.$$

A minimal polynomial is such a polynomial of the smallest degree that $\psi(A) = 0$.

We shall introduce for consideration auxiliary polynomials

$$\psi_k(z) = \frac{\psi(z)}{(z - \lambda_k)^{v_k}} \qquad (k = 1, 2, ..., q). \tag{99}$$

Since polynomials $\psi_k(z)$ are relatively prime then there will be polynomials $f_k(z)$ satisfying identity

$$f_1(z)\psi_1(z) + f_2(z)\psi_2(z) + ... + f_q(z)\psi_q(z) \equiv 1.$$

We shall transfer in identity (98) from scalar argument z to matrix argument A. We shall receive matrix equality

$$f_1(A)\psi_1(A) + f_2(A)\psi_2(A) + ... + f_q(A)\psi_q(A) = E.$$

We shall show that components in equality (99)

$$P_k \equiv f_k(A)\psi_k(A) \tag{100}$$

Are projectors defined by eigenvalues λ_k of matrix .
Since polynomial

$$(z-\lambda_k)^{\nu_k} f_k(z)\psi_k(z) = \psi(z)f_k(z)$$

is divide evenly into minimal polynomial $\psi(z)$, then the following equalities will be true

$$(A-\lambda_k E)^{\nu_k} f_k(A)\psi_k(A) = (\Lambda-\lambda_k E)^{\nu_k} P_k \quad (k=1,2,...,q).$$
(101)

Consequently the rows of matrix P_k enter invariant subspace L_k of matrix A, corresponding eigenvalue λ_k.

Since polynomial $f_k(z)\psi_k(z)f_s(z)\psi_s(z)$ $(k \neq s)$ is divided evenly into a minimal polynomial then

$$f_k(A)\psi_k(A)f_s(z)\psi_s(A) = P_k P_s = 0 \qquad .$$
(102)

Finally, multiplying equality

$$P_1 + P_2 + ... + P_q = E,$$

which was received from (98), by P_k, we receive in the force of formulas (102) equality $P_k P_k = P_k$. It follows from here that matrixes P_k (100) create a full group of projectors.

Notice. If nullifying polynomial   of matrix of m order is divided into two relatively prime multipliers

$$\psi(z) = \psi_1(z)\cdot \psi_2(z),$$
(103)

then the whole   -dimensional space L can correspondingly be expanded into a direct sum of two subspaces L_1, L_2, which are invariant subspaces of matrix A. At that the following expansion will be true

$$X = X_1 + X_2, \quad X_k \in L_k \quad (k = 1,2); \quad \psi_1(A)X_2 = 0,$$
$$\psi_2(A)X_1 = 0. \quad (104)$$

In order to build projectors P_1, P_2 we shall find such polynomials $f_1(z)$, $f_2(z)$ that

$$f_1(z)\psi_1(z) + f_2(z)\psi_2(z) = 1.$$
$$(105)$$

Similarly we can prove that projectors which define expansion (104), look like

$$P_1 = f_1(A)\psi_1 A, \quad P_2 = f_2(A)\psi_2(A).$$
$$(106)$$

Example. We shall find projectors of matrix

$$A = \begin{pmatrix} 1 & 2 \\ 4 & 3 \end{pmatrix}.$$

We shall make a characteristic polynomial which is expanded into two multipliers

$$\begin{vmatrix} z-1 & -2 \\ -4 & z-3 \end{vmatrix} = (z+1)(z-5).$$

Out of identity

$$\frac{1}{6}(z+1) - \frac{1}{6}(z-5) \equiv 1$$

we find projectors of matrix :

$$P_1 = \frac{1}{6}(A + E) = \frac{1}{3}\begin{pmatrix} 1 & 1 \\ 2 & 2 \end{pmatrix},$$

$$P_2 = -\frac{1}{6}(A - 5E) = \frac{1}{3}\begin{pmatrix} 2 & 1 \\ -2 & 1 \end{pmatrix}.$$

Application of the theory of linear projectors for finding eigenvalues of a matrix is shown in works [1, 4]. In work [3] we introduce non-linear projectorsвведены $P_j(t, X)$ $(j = 1,2)$, which satisfy properties:

$$P_j(t, P_j(t, X)) \equiv P_j(t, X), \quad P_j(t, P_s(t, X)) \equiv 0$$

$$(j \neq s, \ j, s = 1,2),$$

$$P_1(t, X) + P_2(t, X) \equiv X, \quad P_j(t, 0) \equiv 0.$$

$$(107)$$

Non-linear projectors enable to split solutions of the non-linear system of differential equations.

12. Application of the matrix theory
to differential and difference equations

The system of linear differential equations

$$\frac{dX}{dt} = AX \tag{108}$$

has a functional matrix of solutions

$$N(t) = e^{At}. \tag{109}$$

Solution of Cauchy problem $t = t_0$, $X = X_0$ looks like имеет

$$X(t) = e^{A(t-t_0)} X(t_0).$$

Example. We shall find a differential matrix of solutions of the system of differential equations (108), where

$$A = \begin{pmatrix} 1 & 2 \\ 4 & 3 \end{pmatrix}, \quad \lambda_1 = -1, \quad \lambda_2 = 5. \tag{110}$$

We find projectors and the general view of the function of matrix A

$$P_1 = \frac{1}{3}\begin{pmatrix} 2 & -1 \\ -2 & 1 \end{pmatrix}, \quad P_2 = \frac{1}{3}\begin{pmatrix} 1 & 1 \\ 2 & 2 \end{pmatrix},$$

$$\varphi(A) = P_1\varphi(\lambda_1) + P_2\varphi(\lambda_2).$$

For function $\varphi(z) = e^{zt}$ we receive

$$e^{At} = \frac{1}{3}\begin{pmatrix} 2 & -1 \\ -2 & 1 \end{pmatrix}e^{-t} + \frac{1}{3}\begin{pmatrix} 1 & 1 \\ 2 & 2 \end{pmatrix}e^{5t} = \begin{pmatrix} \dfrac{e^{5t} + 2e^{-t}}{3} & \dfrac{e^{5t} - e^{-t}}{3} \\ \dfrac{2e^{5t} - 2e^{-t}}{3} & \dfrac{2e^{5t} + e^{-t}}{3} \end{pmatrix},$$

a fundamental matrix of solutions

Similarly we can solve the system of linear difference equations

$$X_{n+1} = AX_n \qquad (n = 0,1,2,...).$$

$$(111)$$

This system of equations has general solution

$$X_n = A^n X_0 \qquad .$$

Example. We shall find the solution of the system of difference equations (111), where matrix A is given in formula (110). For function $\varphi(z) = z^n$ we receive

$$A^n = P_1(-1)^n + P_2(5^n) = \frac{1}{3}\begin{pmatrix} 2 & -1 \\ -2 & 1 \end{pmatrix}(-1)^n + \frac{1}{3}\begin{pmatrix} 1 & 1 \\ 2 & 2 \end{pmatrix}5^n$$

The general solution of the system looks like

$$X_n = \frac{1}{3}\begin{pmatrix} 5^n + 2\cdot(-1)^n & 5^n + (-1)^n \\ 2\cdot 5^n - 2\cdot(-1)^n & 2\cdot 5^n + (-1)^n \end{pmatrix}X_0$$

$$.$$

Similarly we solve systems of equations of the order higher than the first. We look for the solution of the system of differential equations

$$\frac{d^m Y(t)}{dt^m} + \sum_{s=0}^{m-1} A_s \frac{d^s Y(t)}{dt^s} = 0$$

like

$$Y(t) = e^{\lambda t} C.$$

At that we come to the system of equations

$$\left(E\lambda^m + \sum_{s=0}^{m-1} A_s \lambda^s \right) C = 0.$$

Index λ satisfies equation

$$\det\left(E\lambda^m + \sum_{s=0}^{m-1} A_s \lambda^s \right) = 0. \tag{112}$$

For each $\lambda = \lambda_j$ we find corresponding vector C_j. If all roots of equation (112) are different then we find a general solution of the system of differential equations

$$Y(t) = \sum C_j e^{\lambda_j t}.$$

We shall show Valeev K.G.'s result [4] about expansion of polynomial matrixes into multipliers.

Theorem. So that the expansion of polynomial matrixes into multipliers can be true

$$L(\lambda) = L_1(\lambda) L_2(\lambda),$$

where we introduce symbols

$$L(\lambda) = E\lambda^m + \sum_{s=0}^{m-1} A_s \lambda^s, \quad m = p + q, \quad p \geq 1, \quad q \geq 1,$$

$$L_1(\lambda) = E\lambda^p + \sum_{s=0}^{p-1} B_s \lambda^s, \quad L_2(\lambda) = E\lambda^q + \sum_{s=0}^{q-1} C_s \lambda^s,$$

it is necessary and sufficient that any solution of the system of linear differential equations

$$\frac{d^q Y(t)}{dt^q} + \sum_{s=0}^{q-1} C_s \frac{d^s Y(t)}{dt^s} = 0$$

can be solution of the system of differential equations

$$\frac{d^m Y(t)}{dt^m} + \sum_{s=0}^{m-1} A_s \frac{d^s Y(t)}{dt^s} = 0.$$

This theorem can be generalized in the case of time-varying differential operators.

Theorem. So that the linear operator $L(t,d)$

$\left(d \equiv \dfrac{d}{dt} \right)$ can be expanded into multipliers

$$L(t,d) = L_1(t,d) \cdot L_2(t,d),$$

where we introduce symbols

$$L(t,d) = Ed^m + \sum_{s=0}^{m-1} A_s(t) d^s, \quad m = p+q, \quad p \geq 1, \quad q \geq 1,$$

$$L_1(t,d) = Ed^p + \sum_{s=0}^{p-1} B_s(t) d^s,$$

$$L_2(t,d) = Ed^q + \sum_{s=0=}^{q-1} C_s(t) d^s,$$

it is necessary and sufficient that any solution of the system of differential equations

$$\frac{d^q Y(t)}{dt^q} + \sum_{s=0}^{q-1} C_s(t) \cdot \frac{d^s Y(t)}{dt^s} = 0$$

should be the solution of the system of differential equations

$$\frac{d^m Y(t)}{dt^m} + \sum_{s=0}^{m-1} A_s(t) \cdot \frac{d^s Y(t)}{dt^s} = 0.$$

The similar result is true for difference operators [4].

Example. We shall consider linear differential equation with variable coefficients

$$y'' + a_1(t)y' + a_0(t)y = 0.$$

We shall find conditions under which any solution of equation

$$y' + c(t)y = 0$$

will be solution of the equation of the second order. Differentiating equation $y' = -c(t)y$ substituting y', y'' we shall receive equation

$$c'(t) = c^2(t) - a_1(t)c(t) + a_0(t).$$

We shall write the equation of the second order like

$$(d + b(t))(d + c(t))y = 0, \quad d = \frac{d}{dt}.$$

We come to solution

$$y'' + (b(t) + c(t))y' + (c'(t) + b(t)c(t))y = 0.$$

Excluding $b(t)$ out of the system of equations

$$b(t) + c(t) = a_1(t), \quad c'(t) + b(t)c(t) = a_0(t),$$

we shall receive the condition of expanding into multipliers

$$c'(t) = c^2(t) - a_1(t)c(t) + a_0(t).$$

The condition of expanding of a differential operator into multipliers is the condition of existing of a particular solution like $y' + c(t)y = 0$.

13. Basis of the averaging method [5]

In the application the averaging method is often used. To explain this method we should consider the following example.

In Kiev the method of averaging has been developed very actively, there have been a number of publications, conferences and eventually the quantity was transformed into quality. In the publication [5] was explained the main point of the averaging method, namely, the connection of the method of averaging with normal forms of the system of differential equations.

Definition. The system of differential equations

$$\frac{dX}{dt} = AX + F(X) \tag{113}$$

will be called equivalent to the normal form relatively matrix A, if the following identity is fulfilled

$$e^{At}F(e^{-At}X) \equiv F(X), \quad -\infty < t < \infty. \tag{114}$$

Replacement of the variables

$$X = e^{At}Y$$

transforms the system of equations (113) into the system of equations

$$\frac{dY}{dt} = F(X).$$

The given definition of the normal form coincides with the existing definition of the normal form [6].

Let A be a diagonal matrix with eigenvalues $\alpha_1, \alpha_2, ..., \alpha_m$. At that we have equality

$$e^{At} = \begin{pmatrix} e^{\alpha_1 t} & 0 & ... & 0 \\ 0 & e^{\alpha_2 t} & ... & 0 \\ ... & ... & ... & ... \\ 0 & 0 & ... & e^{\alpha_m t} \end{pmatrix}.$$

If we define

$$F(X) = \begin{pmatrix} f_1(x) \\ f_2(x) \\ ... \\ f_m(x) \end{pmatrix}, \tag{115}$$

then from condition (114) we receive

$$f_j(X) = x_j \varphi_j(X) \quad (j = 1, 2, ..., m), \tag{116}$$

where $\varphi_j(X)$ are integrals of the system of equations

$$\frac{dX}{dt} = AX,$$

i.e. identities are fulfilled

$$\varphi_j\left(x_1 e^{\alpha_1 t}, x_2 e^{\alpha_2 t}, ..., x_m e^{\alpha_m t}\right) \equiv \varphi_j\left(x_1, x_2, ..., x_m\right).$$

Example. If the system of differential equations

$$\frac{dx}{dt} = \alpha x + f_1(x, y), \qquad \frac{dy}{dt} = -\alpha y + f_2(x, y), \quad \alpha \neq 0$$

is equivalent to the normal form, then it looks like

$$\frac{dx}{dt} = \alpha x + x\varphi_1(xy), \qquad \frac{dy}{dt} = -\alpha y + y\varphi_2(xy).$$

Transforming to the normal form happens with the help of generalization of the averaging form. Namely, we assume

$$\left\langle t^n e^{\alpha t}\right\rangle = 0 \qquad \left(n = 0,1,2,...; \ \alpha \neq 0\right)$$

(117)

$$\left\langle c\right\rangle = C, \quad c = const$$

$$\left\langle t^n\right\rangle \text{ does not exist for } n = 1,2,3,....$$

If $y(t)$ is a function which has an image by Laplace

$$f(p) = \int_0^\infty e^{-pt} y(t)dt,$$

then we assume [7]

$$\left\langle y(t)\right\rangle = \lim_{p \to 0} pf(p).$$

(118)

We consider the system of differential equations

$$\frac{dZ}{dt} = AZ + \mu F(Z,\mu),$$

$$(119)$$

where vector-function $F(Z,\mu)$ is expanded into formal degree lines by degrees z_j, μ.

After linear replacement

$$Z = e^{At} X \qquad (120)$$

the system of equations (119) is transformed into the non-stationary system of deferential equations

$$\frac{dX}{dt} = \mu e^{-At} F(e^{At} X, \mu),$$

$$(121)$$

which does not have a normal form relatively matrix A.

We have the replacement of variables done

$$X = Y + \mu \Psi(t,Y,\mu); \quad \langle \Psi(t,Y,\mu) \rangle = 0,$$

$$(122)$$

it transforms equations system (121) into a stationary system of equations

$$\frac{dY}{dt} = \mu \Pi(Y,\mu),$$

$$(123)$$

the right part of which does not depend explicitly on time t. Replacement of the variables

$$U = e^{At} Y$$

transforms the system of equations (123) into the system of equations

$$\frac{dU}{dt} = AU + \mu e^{At}\Pi\!\left(e^{-At}U,\mu\right).$$

(124)

Vector-function $\Pi(U,\mu)$ has normal from relatively matrix A, i.e. the following identity is true

$$e^{At}\Pi\!\left(e^{-At}U,\mu\right) \equiv \Pi(U,\mu).$$

Replacement of variables

$$Z = U + \mu e^{At}\Psi\!\left(t, e^{-At}U,\mu\right)$$

does not depend actually on time in the force of identity

$$e^{At}\Psi\!\left(t, e^{-At}U,\mu\right) \equiv \Psi(0,U,\mu).$$

As follows from here, replacement of variables

$$Z = U + \mu\Psi(0,U,\mu)$$

is normalizing, i.e. it transforms the initial system of equations (119) into the system of equations

$$\frac{dU}{dt} = AU + \mu\Pi(U,\mu),$$

(125)

which has a normal form.

The validity of the given statements comes out of the following major theorem [7].

Theorem. If for the system of differential equations

$$\frac{dX}{dt} = \mu e^{-At} F\left(e^{At} X, \mu\right),$$

(126)

where $F(X,\mu)$ is expanded into formal degree lines by degrees $x_1,...,x_m$, μ the replacement is built

$$X = Y + \mu \Psi(t,Y,\mu); \quad \langle \Psi(t,Y,\mu) \rangle \equiv 0,$$

(127)

which makes the system of equations (126) look like

$$\frac{dY}{dt} = \mu \Pi(y,\mu),$$

(128)

then for replacement (127) and equation system (128) the following properties are fulfilled:

$$e^{A\tau} \Pi\left(e^{-A\tau} U, \mu\right) \equiv \Pi(U,\mu);$$
$$e^{A\tau} \Psi\left(t, e^{-A\tau} U, \mu\right) \equiv \Psi(t - \tau, U, \mu).$$

(129)

During the proof of the theorem we use the integrating operation with a zero average.

$$\{Y(t)\} = \int_0^t \left(Y(t) - \langle Y(t) \rangle\right) dt - \left\langle \int_0^t \left(Y(t) - \langle Y(t) \rangle\right) dt \right\rangle,$$

(130)

i.e. the function is averaged both before and after integrating.

Example. We shall find some functions

$$\left\{e^{2t} + 1\right\} = \frac{1}{2} e^{2t}; \quad \left\{\sin^2 t\right\} = \left\{\frac{1 - \cos 2t}{2}\right\} = \frac{-\sin 2t}{4};$$

$$\left\{e^{3t}+e^{4t}+e^{5t}+6\right\}=\frac{1}{3}e^{3t}+\frac{1}{4}e^{4t}+\frac{1}{5}e^{5t};$$

$$\left\{8h^{2}t\right\}=\frac{e^{2t}}{8}-\frac{e^{-2t}}{8}.$$

We shall make replacement (127) in the system of equations (126) and come to the system of equations with partial derivatives of the first order

$$\mu\Pi(Y,\mu)+\mu\frac{\partial\psi(t,Y,\mu)}{\partial t}+\mu^{2}\frac{\mathcal{Д}\Psi(t,Y,\mu)}{\mathcal{Д}Y}\Pi(Y,\mu)=$$

$$=\mu e^{-Et}F\left(e^{At}Y+\mu e^{Et}\Psi(t,Y,\mu),\mu\right).$$

(131)

We use the averaging operation

$$\Pi(Y,\mu)=\left\langle e^{-At}F\left(e^{At}Y+\mu e^{At}\Psi(t,Y,\mu)\right)\right\rangle$$

(132)

and the operation of integrating with averaging

$$\Psi(t,Y,\mu)=\left\langle e^{-At}F\left(e^{At}Y+\mu e^{At}\Psi(t,Y,\mu),\mu\right)-\mu\frac{D\psi(t,Y,\mu)}{DY}\Pi(Y,\mu)\right\rangle$$

(133)

The solution of system (132), (133) can be found by using expansions by degrees of parameter μ of by the successive approximation method. While using the successive approximation method for

$$\Pi_{0}(Y,\mu)\equiv 0,\quad \psi_{0}(t,Y,\mu)\equiv 0$$

we receive

$$\Pi_{k+1}(Y,\mu)=\left\langle e^{-At}F\left(e^{At}Y+\mu e^{At}\Psi_k(t,Y,\mu),\mu\right)\right\rangle;$$

$$\Psi_{k+1}(t,Y,\mu)=\left\{e^{-At}F\left(e^{At}Y+\mu e^{At}\Psi_k(t,Y,\mu),\mu\right)-\right.$$

$$\left.-\mu\frac{\partial\Psi_k(t,Y,\mu)}{\partial Y}\Pi_k(Y,\mu)\right\}\quad(k=0,1,2,\ldots).$$

$$(134)$$

At that we receive asymptotic relations

$$\left\|\Pi_k(Y,\mu)-\Pi(Y,\mu)\right\|=O(\mu^k);$$

$$\left\|\Psi_k(t,Y,\mu)-\Psi(t,Y,\mu)\right\|=O(\mu^k)\quad(k=0,1,2,\ldots),$$

$$(135)$$

i.e. the first parts of expansion of vector-functions $\Pi(Y,\mu)$, $\Psi(t,Y,\mu)$ by degree μ can be received with precision to within μ^k inclusively from expansions by degree μ of approximations $\Pi_k(Y,\mu)$, $\Psi_k(t,Y,\mu)$.

We shall consider the system of differential equations

$$\frac{dZ_1}{dt}=A_1Z_1+\mu F_1(Z_1,Z_2,\mu);$$

$$\frac{dZ_2}{dt}=A_2Z_2+\mu F_2(Z_1,Z_2,\mu),\quad F_k(0,0,\mu)\equiv0\quad(k=1,2).$$

$$(136)$$

We assume that vector-functions $F_k(Z_1,Z_2,\mu)$ are differentiated by all arguments as many times as necessary.

Let eigenvalues of matrix A_1 lie on an imaginary axis and matrix A_1 have a simple structure. Eigenvalues of matrix A_2 have negative real parts.

Using the proposed asymptotic method we transform the system of equations (136) into the system of equations which looks like

$$\frac{dU_1}{dt} = A_1 U_1 + \mu \Pi (U_1, 0, \mu),$$

$$\frac{dU_2}{dt} = A_2 U_2 + \mu \Pi (U_1, U_2, \mu). \quad (137)$$

Critical variables are separated into another system of equations. Stability of the zero solution of system (136) is equal to stability of the zero solution of first system (137).

Example. We shall study stability of the zero solution of the system of differential equations

$$\frac{dZ_1}{dt} = \mu Z_2 + \mu^2 a Z_1^3, \quad \frac{dZ_2}{dt} = -Z_2 + \mu b Z_1^3.$$

$$(138)$$

The system of equations (138) is converged to one equation

$$\frac{dU_1}{dt} = \mu^2 (a+b) U_1^3 + O(\mu^3 U_1^4).$$

So the system of equations (138) has a steady zero solution for $a+b<0$ and a non-steady one for $a+b>0$.

The proposed new asymptotic method allows to separate critical variables.

14. Matrix spectrum expansion

In conclusion we shall explain some not very well-known results about the matrix spectrum expansion.

The spectrum expansion of matrix Λ of m-order is supposed to be a creation of two matrixes Λ_1, Λ_2 of such orders q, $p = m - q$ that the spectrum of matrix is a sum of the spectrums of matrixes , . The spectrum of matrix , i.e. a set of all eigenvalues of matrix , is defined as $S_r\Lambda$. In the general case expansion of the spectrum of matrix is called a creation of matrixes $\Lambda_1, \Lambda_2, ..., \Lambda_r$ of a lower order than and such as

$$S_r\Lambda_1 \bigcup \qquad \bigcup \qquad S_r\Lambda, \tilde{a}\ddot{a}\mathring{a}$$
$$S_r-\tilde{n}\ddot{u}\;\mathring{a}\hat{e}\grave{o}\;\eth\grave{i}\;\grave{a}\grave{o}\;\eth\grave{e}\ddot{o}\hat{u}$$

Expansion of the matrix spectrum is applied in particular in the case when it is necessary to find only parts of the spectrum lying in the given area. The matrix expansion can be fulfilled in different ways but we shall explain the simplest and the most accessible ones.

Let us have some square matrix

$$\Lambda = \begin{pmatrix} A & B \\ C & D \end{pmatrix} \tag{139}$$

of $m \times m$ size, which is divided into blocks. Here matrix A has size $q \times q$, matrix D has size $p \times p$ ($p = m - q$, $q \geq 1$, $p \geq 1$).

Suppose we found a transformation like

$$T^{-1} \Lambda T = \begin{pmatrix} \Lambda_{11} & 0 \\ 0 & \Lambda_{22} \end{pmatrix}, \quad \dim \Lambda_{11} = q \times q, \quad \dim \Lambda_{22} = p.$$

At that the spectrum of matrix Λ is found with the help of spectrums of matrixes $\Lambda_{11}, \Lambda_{22}$

$$S_r \Lambda = S_r \Lambda_{11} \bigcup S_r \Lambda_{22}.$$

The similar result is valid for transformations of similarities of matrix into block matrixes

$$T_1^{-1} \Lambda T_1 = \begin{pmatrix} \Lambda_{11} & \Lambda_{12} \\ 0 & \Lambda_{22} \end{pmatrix}, \quad T_2^{-1} \Lambda T_2 = \begin{pmatrix} \Lambda_{11} & 0 \\ \Lambda_{21} & \Lambda_{22} \end{pmatrix}.$$

In order to build the transformation we use a creation of the integral varieties of the systems of differential or difference equations.

Differential equation of expansion

The ways to create integral varieties for the system of differential equations are applied to find eigenvalues and invariant sub-spaces of matrixes.

For the system of differential equations

$$\frac{dX}{dt} = AX + BY, \quad \frac{dY}{dt} = CX + DY$$

$$(140)$$

we find an integral variety of solutions

$$Y = K(t)X,\qquad(141)$$

where $K(t)$ is a matrix of $p \times q$ size. Differentiating vector equation (141) by t in force of the initial system of equations (140), we come to equality

$$CX + DY = \frac{dK(t)}{dt}X + K(t)(AX + BY).$$

$$(142)$$

If equality (142) is fulfilled in force of the system of equations (141), then the system of equations (141) defines the integral variety of solutions for system (140). Excluding from the system of equations (142) Y, we come to matrix differential equation

$$\frac{dK(t)}{dt} = C + DK(t) - K(t)(A - BK(t)),$$

$$(143)$$

which is called the differential equation of expansion. The stationary solution of matrix equation (143) satisfies the matrix equation of expansion

$$C + DK - KA - KBK = 0.$$

$$(144)$$

Notice. In the theory of optimal controlling for the synthesis of an optimal regulator for the special system of differential equations

$$\frac{dX}{dt} = AX - MY, \quad \frac{dY}{dt} = -QX - A^*Y \qquad (\dim X = \dim Y),$$

$$(145)$$

which defines the necessary conditions of optimality, matrix equation (143) for matrix $K(t)$ looks like

$$\frac{dK(t)}{dt} = -Q - A^*K(t) - K(t)A + K(t)MK(t)$$

$$(146)$$

and it is often **called the matrix differential equation of Riccati** in the theory of controlling. The stationary solution of equation (146) satisfies the matrix equation

$$-Q - A^*K - KA + KMK = 0,$$

$$(147)$$

which is called the Riccati equation [8].

Below it is demonstrated that the systems of equations (146), (147) are closely connected with expansion of the matrix spectrum.

Let L_1 be some invariant subspace of matrix Λ (139) of q dimension, shown like

$$\begin{pmatrix} X \\ Y \end{pmatrix} = \alpha_1 \begin{pmatrix} X_1 \\ Y_1 \end{pmatrix} + \alpha_2 \begin{pmatrix} X_2 \\ Y_2 \end{pmatrix} + ... + \alpha_q \begin{pmatrix} X_q \\ Y_q \end{pmatrix},$$

$$(148)$$

where $\alpha_1, \alpha_2 ..., \alpha_q$ are scalar coefficients of expansion.

We shall compose auxiliary matrixes from basis vectors

$$T_{11} = (X_1 X_2 ... X_q), \ T_{21} = (Y_1 Y_2 ... Y_q).$$

Vector equalities (148) can be written like

$$X = T_{11}\alpha, \ Y = T_{21}\alpha, \ \alpha = \begin{pmatrix} \alpha_1 \\ ... \\ \alpha_q \end{pmatrix}.$$

$$(149)$$

If $\det T_{11} \neq 0$, then excluding coefficients $\alpha_1, ..., \alpha_q$ from the system of equations (149), we can write subspace L_1 like the system of equations

$$Y = KX, \ K = T_{21} T_{11}^{-1}. \tag{150}$$

Theorem. If matrix Λ (139) has invariant subspace L of q dimension, represented by the system of equations (150), then matrix K of $p \times q$ size is the solution of expansion equation (144).

Theorem. If expansion equation (144) has some , then the columns of matrix

$$Q_1 = \begin{pmatrix} E_q \\ K \end{pmatrix}$$

of $m \times q$ size define the invariant subspace of dimension of matrix   .

Proof. The validity of the theorem comes out of equality

$$\begin{pmatrix} A & B \\ C & D \end{pmatrix} \begin{pmatrix} E_q \\ K \end{pmatrix} = \begin{pmatrix} E_q \\ k \end{pmatrix} (A + BK),$$

which is equal to expansion equation (144).

Here E_q is a unit matrix of q-order.

Theorem. If the equation of expansion (144) has some solution K of $p \times q$ size, then the lines of matrix

$$R_2 = \left(-K \vdots E_p\right) \tag{151}$$

define invariant subspace L_2^* of p dimension of matrix Λ in the conjugate space.

The validity of the theorem comes out of equality

$$\left(D - KB\right)\left(-K \vdots E_p\right) = \left(-K \vdots E_p\right)\begin{pmatrix} A & B \\ C & D \end{pmatrix}, \tag{152}$$

equal to equality (144).

Let the found matrix be the solution of matrix equation (144).

We shall introduce two reciprocal block matrixes

$$T_1 = \begin{pmatrix} E_q & 0 \\ K & E_p \end{pmatrix}, \qquad T_1^{-1} = \begin{pmatrix} E_q & 0 \\ -K & E_p \end{pmatrix} \tag{153}$$

and use the transformation of similarity

$$T_1^{-1} \Lambda T_1 = \begin{pmatrix} A + BK & B \\ C + DK - K(A + BK) & D - KB \end{pmatrix}, \tag{154}$$

which does not change eigenvalues of the matrix.

If matrix K is the solution of the equation of expansion (144), then out of formula (154) we receive equality

$$T_1^{-1} \Lambda T_1 = \begin{pmatrix} A+BK & B \\ 0 & D-KB \end{pmatrix}. \tag{155}$$

Consequently the spectrum of matrix Λ can be expanded into spectrums of matrixes $A+BK$, $D-KB$, i.e.

$$S_r \Lambda = S_r(A+BK) \cup S_r(D-KB).$$

Thus, finding eigenvalues of matrix   of m order can be reduced to finding eigenvalues of two matrixes of q, p $(q+p=m)$ order.

Similarly we find an integral variety of solutions of the system of equations (140), defined by vector equation

$$X = S(t)Y, \tag{156}$$

where $S(t)$ is a matrix of $q \times p$ size. For matrix   we find the following differential equation of expansion

$$\frac{dS(t)}{dt} = B + AS(t) - S(t)D - S(t)CS(t).$$

$$\tag{157}$$

The stationary solution of equation (157) satisfies the matrix equation of expansion

$$B + AS - SD - SCS = 0.$$

$$(158)$$

As stated above, the following theorem is true.

Theorem. So that matrix Λ (139) can have invariant subspace L_2 of $p = m - q$ dimension with the basis which is defined by columns $m \times p$ of matrix

$$Q_2 = \begin{pmatrix} S \\ E_p \end{pmatrix},$$

$$(159)$$

it is necessary and sufficient if $q \times p$ of matrix S satisfies the equation of expansion (158). At that the line of matrix

$$R_1 = \left(E_q \vdots -S \right)$$

$$(160)$$

defines the invariant subspace L_1 of q dimension of matrix   in the conjugate space.

For the expansion of matrix   we can use block matrixes

$$T_2 = \begin{pmatrix} E_q & S \\ 0 & E_p \end{pmatrix}, \quad T_2^{-1} = \begin{pmatrix} E_q & -S \\ 0 & E_p \end{pmatrix}.$$

$$(161)$$

If matrix   satisfies the equation of expansion (158), then we receive the following equality

$$T_2^{-1} \Lambda T_2 = \begin{pmatrix} A - SC & 0 \\ C & CS + D \end{pmatrix}.$$

$$(162)$$

Consequently finding the spectrum of matrix Λ can be reduced to finding spectrums of matrixes $A-SC$, $CS+D$, i.e.

$$S_r\Lambda = S_r(A-SC)\cup S_r(CS+D).$$

(163)

We shall give a simple example to demonstrate the results.

Example. We shall consider matrix

$$\Lambda = \begin{pmatrix} 1 & 2 \\ 4 & 3 \end{pmatrix},$$

(164)

which has eigenvalues $z_1 = -1$, $z_2 = 5$.

Comparing with (139), we find values

$$A=1,\ B=2,\ C=4,\ D=3.$$

The equation of expansion (144) looks like

$$4+3k-k\cdot1-k2k=0$$

and has two solutions $k_1 = -1$, $k_2 = 2$.

Matrix (164) is transformed into matrix (155)

$$\begin{pmatrix} A+BK & B \\ 0 & D-KB \end{pmatrix} = \begin{pmatrix} 1+2k & 2 \\ 0 & 3-2k \end{pmatrix},$$

which for　　　looks like

$$\begin{pmatrix} 1+2k_1 & 2 \\ 0 & 3-2k_1 \end{pmatrix} = \begin{pmatrix} -1 & 2 \\ 0 & 5 \end{pmatrix},$$

and for　　　it looks like

$$\begin{pmatrix} 1+2k_2 & 2 \\ 0 & 3-2k_2 \end{pmatrix} = \begin{pmatrix} 5 & 2 \\ 0 & -1 \end{pmatrix}.$$

These matrixes have eigenvalues $z_1 = -1$, $z_2 = 5$ on the diagonal.

Similarly expansion equation (158) looks like

$$2 + 1 \cdot S - S \cdot 3 - S \cdot 4 \cdot S = 0$$

and has two solutions $S_1 = -1$, $S_2 = \dfrac{1}{2}$.

Matrix (162) looks like

$$\begin{pmatrix} A-SC & 0 \\ C & CS+D \end{pmatrix} = \begin{pmatrix} 1-4s & 0 \\ 4 & 4s+3 \end{pmatrix}$$

and for $s_1 = -1$ looks like

$$\begin{pmatrix} 1-4s_1 & 0 \\ 4 & 4s_1+3 \end{pmatrix} = \begin{pmatrix} 5 & 0 \\ 4 & -1 \end{pmatrix},$$

and for $s_2 = \dfrac{1}{2}$ looks like

$$\begin{pmatrix} 1-4s_2 & 0 \\ 4 & 4s_2+3 \end{pmatrix} = \begin{pmatrix} -1 & 0 \\ 4 & 5 \end{pmatrix}.$$

These matrixes have eigenvalues   ,   on the diagonal.

15. The solution of expansion equations

In order to find solutions of matrix equations of expansion (147), (158) the asymptotically stable solutions of differential equations of expansion (146), (157) can be used. The following theorem lies in the basis of the solution method.

Theorem. Let spectrum σ of matrix Λ be divided into two non-empty subsets σ_1, σ_2 of some straight $\operatorname{Re} z = \gamma$ so that q eigenvalues of matrix   lie in half-plane $\operatorname{Re} z > \gamma$, and these eigenvalues create set   , and the corresponding invariant subspace is found by vector equation $Y = K_2 X$. Let there in the half-plane $\operatorname{Re} z < \gamma$ be $p = m - q$ eigenvalues of matrix   , all of which create set   , and the corresponding invariant subspace is found by equation $X = S_1 Y$. Then matrix K_2 is the single asymptotically stable for $t \to +\infty$ solution of the differential equation of expansion

$$\frac{dK(t)}{dt} = C + DK(t) - K(t)A - K(t)BK(t).$$

(165)

Matrix S_1 is the single asymptotically stable for $t \to -\infty$ solution of the differential equation of expansion

$$\frac{dS(t)}{dt} = B + AS(t) - S(t)D - S(t)CS(t).$$

(166)

The following theorem is similar to the previous one.

Theorem. Let spectrum σ of matrix Λ be divided into two non-empty sets σ_1, σ_2 of straight $\mathrm{Re} z = \gamma$ so that in half-plane $\mathrm{Re} z < \gamma$ there is set σ_1, and in half-plane $\mathrm{Re} z > \gamma$ there is set σ_2.

Let the invariant subspace of matrix , corresponding to , be found by equation $Y = K_1 X$, and the invariant subspace, corresponding to , be found by equation $X = S_2 Y$. Then the differential equation of expansion (165) has the single asymptotically stable for $t \to -\infty$ solution $K(t) = K$, and differential equation (166) has the single asymptotically stable for $t \to +\infty$ solution S_2.

Example. We shall consider matrix (164). Differential equations of expansion (165), (166) will look like

$$\frac{dk(t)}{dt} = 4 + 3k(t) - k(t) \cdot 1 - k(t) \cdot 2k(t),$$

$$(167)$$

$$\frac{ds(t)}{dt} = 2 + 1 \cdot s(t) - s(t) \cdot 3 - s(t) 4 s(t).$$

$$(168)$$

The spectrum of matrix (164) can be divided by straight $\mathrm{Re} z = 0$ so that for $\mathrm{Re} z < 0$ there is number $\sigma_1 = -1$, and for $\mathrm{Re} z > 0$ there is number $\sigma_2 = 5$.

Differential equation (167) can be re-written like

$$\frac{dk(t)}{dt} = -2(k(t)-2)(k(t)+1).$$

Integral curves are shown in fig.1.

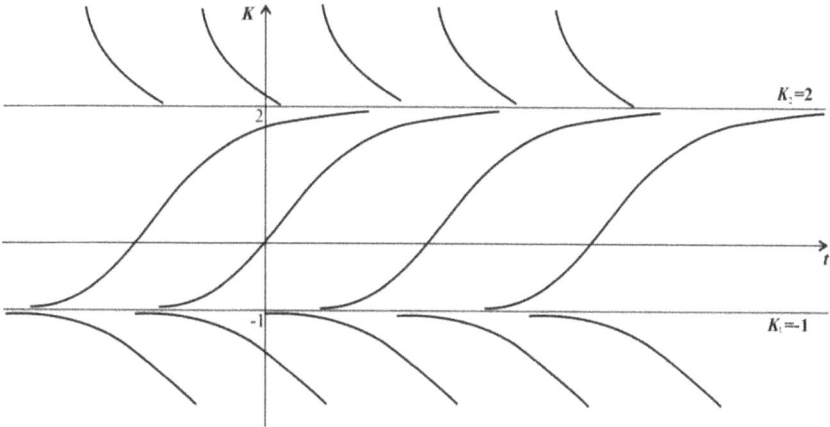

Fig. 1.

There is one asymptotically stable for $t \to +\infty$ solution $k = 2$. Another solution $k = -1$ is asymptotically stable for $t \to -\infty$.

Equation (168) can be written like

$$\frac{ds(t)}{dt} = -2\left(s(t)-\frac{1}{2}\right)(s(t)+1).$$

This equation has the single asymptotically stable for $\quad$ solution $s = \frac{1}{2}$ and the asymptotically stable for $\quad$ solution $s = -1$. The integral curves are shown in fig. 2.

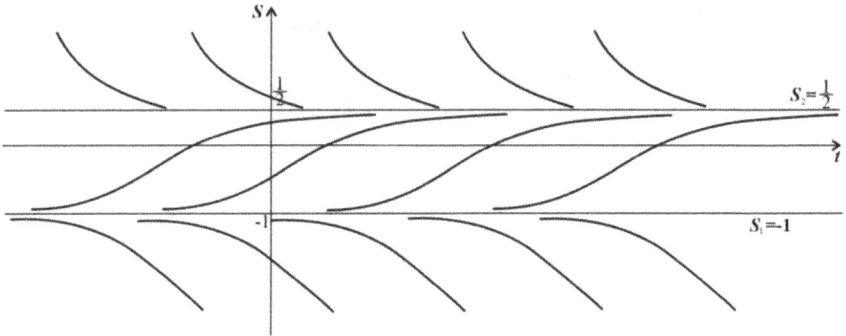

Fig. 2.

Thus in order to find the solution of the expansion equation

$$4+3k-k\cdot1-k2k=0,$$

$$2+1\cdot s-s\cdot3-s\cdot4s=0$$

we need to integrate numerically differential equations of expansion (167), (168) for $t\to+\infty$ or for $t\to-\infty$ and find asymptotically stable solutions which allow to expand the spectrum of matrix Λ. Since an overfill is possible, then for big enough by norm values $\|K(t)\|$ in equation (143) or big enough by norm values $\|S(t)\|$ in equation (157) we need to stop the calculations and choose randomly new initial values $K(0)$, $S(0)$. For matrix   of high order the choice of initial values   , becomes a difficult problem.

Note. While studying the stability of solutions of the system of linear differential equations

$$\frac{dZ}{dt} = \Lambda(\mu)Z, \tag{169}$$

where matrix $\Lambda(\mu)$ has high enough order, it is not necessary to find all eigenvalues of matrix $\Lambda(\mu)$. Using the previous results it is necessary to find one or two eigenvalues of matrix with the biggest real parts. Changing constantly parameter μ we can define the boundaries of the area of stability.

Difference equations of expansion

In order to find invariant sub-spaces of the matrix and for expansion of the spectrum of the matrix we use an integral variety of the system of difference equations.

In order to find invariant sub-spaces of block matrix

$$\Lambda = \begin{pmatrix} A & B \\ C & D \end{pmatrix}$$

we use the following system of difference equations

$$X_{n+1} = AX_n + BY_n, \qquad Y_{n+1} = CX_n + BY_n. \tag{170}$$

We find an integral variety of solutions of the system of difference equations (170) like the system of difference equations

$$Y_n = K_n X_n \qquad (n = 0, \pm 1, \pm 2, \ldots), \tag{171}$$

where K_n is a matrix of $p \times q$ $(p+q=m)$ size.

Replacing n by $n+1$ we receive the following system of equations

$$Y_{n+1} = K_{n+1} X_{n+1}.$$

Excluding Y_{n+1}, X_{n+1}, we come to the following equation

$$CX_n + DY_n = K_{n+1}(AX_n + BY_n).$$

(172)

Since equation (172) must be fulfilled along with equation (171), we come to matrix equation

$$C + DK_n = K_{n+1}(A + BK_n),$$

(173)

which is called a difference equation of expansion. For $K_{n+1} = K = const$ equation

$$C + DK = K(A + BK)$$ (174)

coincides with the equation of expansion (144).

The solution of expansion equation (174) can be found by integrating matrix difference equations

$$K_{n+1} = (C + DK_n)(A + BK_n)^{-1} \qquad (n = 0,1,2,...),$$

$$K_n = (K_{n+1}B - D)^{-1}(C - K_{n+1}A) \qquad (n = -1,-2,-3,...).$$

(175)

Similarly it is possible to find integral variety of solutions of difference equations (170) like

$$X_n = S_n Y_n \qquad (n = 0,\pm 1,\pm 2,...).$$

(176)

Matrix S_n of $q \times p$ size satisfies the matrix difference equation of expansion

$$AS_n + B = S_{n+1}(CS_n + D), \tag{177}$$

which for $S_n = S_{n+1} = const$ is transformed into expansion equation (158). The stationary solution of the matrix difference equation of expansion (158) can be found like a limiting solution of difference equations

$$S_{n+1} = (AS_r + B)(CS_n + D)^{-1} \quad (n = 0,1,2,...);$$

$$S_n = (\Lambda - S_{n+1}C)^{-1}(S_{n+1}D - B) \quad (n = -1,-2,-3,...). \tag{178}$$

The following theorems are valid for this.

Theorem. Let spectrum σ of matrix Λ be expanded into two non-empty sets with circle $|z| = \rho > 0$. Let set σ_1 lie in area $|z| < \rho$ and the corresponding invariant subspace of matrix be found by vector equation $Y = K^{(1)}X$, where $K^{(1)}$ is a matrix of $p \times q$. Let σ_2 lie in area $|z| > \rho$ and the corresponding invariant subspace is found by equation $X = S^{(2)}Y$, where $S^{(2)}$ is a matrix of size. Then for almost all initial values of matrixes K_0, S_0 there are asymptotically stable solutions of the system of difference equations of expansion (175), (178) and

$$\lim_{n \to -\infty} K_n = K^{(1)}, \ \lim_{n \to +\infty} S_n = S^{(1)}.$$

$$(179)$$

At that the spectrum of matrix Λ can be expanded and

$$\sigma_1 = S_r\left(A + BK^{(1)}\right) = S_r\left(A - S^{(2)}C\right),$$
$$\sigma_2 = S_r\left(D - K^{(1)}B\right) = S_r\left(D + CS^{(2)}\right).$$

$$(180)$$

Similarly the following result is valid.

Theorem. Let spectrum σ of matrix   be divided into two non-empty sets σ_1, σ_2 of circle $|z| = \rho > 0$. Let σ_2 lie in area $|z| > \rho$ and the corresponding invariant subspace of matrix   can be found by equation $Y = K^{(2)}X$, where $K^{(2)}$ is a matrix of $p \times q$ size. Let σ_1 lie in circle $|z| < \rho$ and the corresponding invariant subspace of matrix   can be found by equation $X = S^{(1)}Y$. Then for almost all initial values K_0, S_0 there is an asymptotically stable stationary solution of the system of difference equations of expansion (175), (178) and there are limits

$$\lim_{n \to +\infty} K_n = K^{(2)}, \qquad \lim_{n \to -\infty} S_n = S^{(1)}.$$

$$(181)$$

For the known matrixes   , $S^{(1)}$ the spectrum of matrix   is expanded and

$$\sigma_1 = S_r\left(D - K^{(2)}B\right) = S_r\left(D + CS^{(1)}\right),$$
$$\sigma_2 = S_r\left(A + BK^{(2)}\right) = S_r\left(A - S^{(1)}C\right).$$

(182)

Example. We shall consider the following matrix

$$\Lambda = \begin{pmatrix} 1 & 2 \\ 4 & 3 \end{pmatrix}, \quad A = 1, \ B = 2, \ C = 4, \ D = 3.$$

(183)

Difference equations of expansion (175) look like

$$K_{n+1} = \frac{4 + 3K_n}{1 + 2K_n} \qquad (n = 0,1,2,\ldots),$$

$$K_n = \frac{4 - K_{n+1}}{2K_{n+1} - 3} \qquad (n = -1,-2,-3,\ldots).$$

(184)

Difference equations of expansion (178) look like

$$S_{n+1} = \frac{S_n + 2}{4S_n + 3},$$

$$S_n = \frac{3S_{n+1} - 2}{1 - 4S_{n+1}}.$$

(183)

Calculations by these formulas are shown in the table for $K_0 = 0$, $S_0 = 0$

n	1	3	5	7	∞
K_n	4,060066	2,048780	2,001921	2,000076	$2 = K^{(2)}$
K_{-n}	-1,333333	-1,012048	-1,000480	-1,000019	$-1 = K^{(1)}$
S_n	0,666667	0,506624	0,500240	0,499952	$0,5 = S^{(2)}$
S_{-n}	-2,000000	-1,024390	-0,995215	-0,999808	$-1 = S^{(1)}$

By formulas (180) we find the spectrum of matrix Λ

$$\sigma_1 = 1 + 2(-1) = 1 - 0,5 \cdot 4 = -1,$$
$$\sigma_2 = 3 - (-1)2 = 3 + 4 \cdot 0,5 = 5.$$

Similarly by formulas (182) we find spectrums

$$\sigma_1 = 3 - 2 \cdot 2 = 3 + 4(-1) = -1,$$
$$\sigma_2 = 1 + 2 \cdot 2 = 1 - (-1) \cdot 4 = 5.$$

Note. In order to find matrixes , by formulas (175), (178) there is usually no overfill of the arithmetic unit.

16. Expansion of polynomial matrixes into matrix multipliers

We shall show some results of expansion of differential and difference matrix operators into matrix multipliers.

Theorem. In order to expand the matrix differential operator

$$L(t,D) \equiv ED^N + \sum_{k=1}^{N} A_k(t)D^{N-k}, \qquad D \equiv \frac{d}{dt}, \quad N \geq 2$$

$$(186)$$

into operator multipliers

$$L(t,D) = L_1(t,D)L_2(t,D),$$

$$L_1(t,D) \equiv ED^p + \sum_{k=1}^{p} B_k(t)D^{p-k},$$

$$L_2(t,D) \equiv ED^q + \sum_{k=1}^{q} C_k(t)D^{q-m}, \qquad p+q = N,$$

$$p \geq 1, q \geq 1$$

it is necessary and sufficient if any solution of the system of differential equations

$$L_2(t,D)Y(t) = 0$$

is also the solution of the system of differential equations

$$L(t,D)Y(t) = 0.$$

Here matrixes $A_k(t)$, $B_k(t)$, $C_k(t)$ are continuous and differentiated by t as many times as necessary.

Example. We shall consider a differential equation of the second order

$$y''(t) + a_1(t)y'(t) + a_2(t)y(t) = 0,$$

which can be written like

$$\left(D^2 + a_1(t)D + a_2(t)y(t)\right) = 0, \quad D \equiv \frac{d}{dt}.$$

(187)

We shall expand the differential operator

$$L(t,D) \equiv D^2 + a_1(t)D + a_2(t)$$

into multipliers

$$D^2 + a_1(t)D + a_2(t) = \left(D + b(t)\right)\left(D + C(t)\right).$$

At that we receive an equation for $C(t)$

$$a_2(t) = C'(t) + a_1(t)C(t) - C^2(t).$$

(188)

On the other side let any solution of the differential equation

$$y' + C(t)y = 0$$

be the solution of equation (187). At that equation (189) is fulfilled as well.

Example. The differential equation of the second order

$$y''(t) + y(t) = 0$$

has a particular solution $y(t) = \sin t$. The differential equation of the first order

$$y'(t) + c(t)y(t) = 0, \quad c(t) = -ctgt$$

also has a solution $y(t) = -\sin t$. It follows thence that the differential operator $D^2 + 1$ can be expanded into multipliers

$$D^2 + 1 = (D + b(t))(D - ctgt).$$

At that we come to the system of equations

$$b(t) - ctgt = 0, \qquad \frac{1}{\sin^2 t} - b(t)ctgt = 1,$$

which is satisfied at $b(t) = ctgt$.

Similar results are valid for difference operators.

We shall denote S a shift operator, i.e.

$$S'Y(t) \equiv Y(t + h), \ h > 0.$$

Theorem. In order to expand the matrix difference operator

$$L(t, S) \equiv ES^N + \sum_{k=1}^{N} A_k(t)S^{N-k} \qquad (189)$$

into operator matrix multipliers

$$L(t, S) = L_1(t, S) \cdot L_2(t, S),$$

$$L_1(t, S) \equiv ES^p + \sum_{k=1}^{p} B_k(t)S^{p-k},$$

$$L_2(t, S) \equiv ES^q + \sum_{k=1}^{q} C_k(t)S^{q-k}, \ p + k = N, \ p \geq 1, \ q \geq 1,$$

it is necessary and sufficient that any solution of the system of difference equations

$$L_2(t, S)Y(t) = 0$$

should be the solution of the system of difference equations

$$L(t,S)Y(t) = 0.$$

Here matrixes $A_k(t)$, $B_k(t)$, $C_k(t)$ are continuous matrixes of the same order.

Example. We shall consider the following difference equation of the second order

$$y(t+2h) + a_1(t)y(t+h) + a_2(t)y(t) = 0,$$

$$(190)$$

which can be written like an operator

$$\left(S^2 + a_1(t)S + a_2(t)\right)y(t) = 0, \qquad Sy(t) \equiv y(t+h).$$

We shall write the difference operator like the product

$$S^2 + a_1(t)S + a_2(t) = \left(S + b(t)\right)\!\left(S + c(t)\right),$$

which is valid at fulfilling the equalities

$$a_1(t) = c(t+h) + b(t), \qquad a_2(t) = b(t)c(t),$$

which implies the equality for $c(t)$

$$c(t)c(t+h) - a_1(t)c(t) + a_2(t) = 0.$$

$$(191)$$

On the other side we shall find a solution at fulfilling which any solution of the difference equation

$$\left(S + c(t)\right)y(t) = 0, \qquad y(t+h) + c(t)y(t) = 0$$

is the solution of equation (190) as well. We shall exclude $y(t+h)$, $y(t+2h)$ from the equations

$$, \qquad y(t+2h) + c(t+h)y(t+h) = 0,$$

$$y(t+2h)+a_1(t)y(t+h)+a_2(t)y(t)=0,$$

and receive the same equation (191).

The case when differential operator (186) does not depend on time t is more important. At that we receive the following important result.

Theorem. In order to expand the matrix differential operator

$$L(D)=ED^N+\sum_{k=1}^{N}A_kD^{N-k}, \ D\equiv\frac{d}{dt}, \ N\geq2,$$

into operator multipliers

$$L(D)=L_1(D)L_2(D),$$

$$L_1(D)\equiv ED^p+\sum_{k=1}^{p}B_kD^{p-k},$$

$$L_2(D)\equiv ED^q+\sum_{k=1}^{q}C_kD^{q-k} \quad (p+q=N, \ p\geq1, \ q\geq1)$$

it is necessary and sufficient that any solution of the system of differential equations

$$L_2(D)Y(t)=0$$

should also be the solution of the system of differential equations

$$L(D)Y(t)=0.$$

Here matrixes $A_k(t)$, $B_k(t)$, $C_k(t)$ have the same size $m\times m$.

The differentiation operator D can be regarded here like a parameter.

Example. In order to divide polynomial $f(x)$ into binomial $(x-a)$ it is necessary and sufficient that $f(a)=0$ (The remainder theorem).

Example. The polynomial of the third order

$$f(x)=x^3-x^2-x+1$$

is divided into polynomial $f_2(x)=(x-1)^2$, if any solution of the differential equation

$$y''-2y'+y=0,$$

i.e. $y_0(t)=(c_1+c_2t)e^t$, is the solution of the differential equation

$$y'''-y''-y'+y=0.$$

$$(192)$$

Since function $y=y_0(t)$ satisfies differential equation (192) for any values of constants c_1,c_2, then polynomial   is divided into polynomial $f_2(x)$.

The previous theorem implies the criterion of divisibility of polynomial matrixes.

Let us have a polynomial matrix

$$L(z)=Ez^N+\sum_{k=1}^{N}A_kz^{N-k}.$$

$$(193)$$

This matrix is divided on the right into $Ez-C$, i.e.

$$L(z)=\left(Ez^{N-1}+\sum_{k=1}^{N-1}B_kz^{N-1-k}\right)(Ez-C),$$

$$(194)$$

if any solution of the equation system

$$(Ez - C)X = 0 \qquad (195)$$

is the solution of the system of equations $L(z)X = 0$. Consequently number $z = z_k$ is an eigenvalue and vector X is the eigenvalue of matrix C. It implies the method of expansion of matrix $L(z)$ into multipliers.

1. We find eigenvalues $z_1, ..., z_{N \cdot m}$ of the bunch of matrixes $L(z)$ from equation $\det L(z) = 0$.

2. We find eigenvector $X_1, ..., X_{Nm}$ of bunch $L(z)$ from the system of equations $L(z_k)X_k = 0$.

3. We chose linearly independent eigenvectors $X_1, ..., X_n$ and recreate matrix   knowing eigenvalues $z_1, ..., z_n$ and corresponding eigenvectors $X_1, ..., X_n$ with the help of projectors.

At that expansion (194) will be valid.

Example. We shall consider polynomial matrix

$$L(z) = \begin{pmatrix} z^2 - 2z + 3 & 3z - 3 \\ 5z - 1 & z^2 + 5 \end{pmatrix}.$$

We find eigenvalues of matrix $L(z)$ from equation $\det L(z) = 0$ and corresponding eigenvector

$$z_1 = 2, \; L(2) = \begin{pmatrix} 3 & 3 \\ 9 & 9 \end{pmatrix}, \; X_1 = \begin{pmatrix} 1 \\ -1 \end{pmatrix};$$

$$z_2 = -2, \; L(-2) = \begin{pmatrix} 11 & -9 \\ -11 & 9 \end{pmatrix}, \; X_2 = \begin{pmatrix} 9 \\ 11 \end{pmatrix};$$

$$z_3 = -1, \; L(-1) = \begin{pmatrix} 6 & -6 \\ -6 & 6 \end{pmatrix}, \; X_3 = \begin{pmatrix} 1 \\ 1 \end{pmatrix};$$

$$z_4 = 3, \; L(3) = \begin{pmatrix} 6 & 6 \\ 14 & 14 \end{pmatrix}, \; X_4 = \begin{pmatrix} 1 \\ -1 \end{pmatrix}.$$

We find matrix C, which has eigenvalues ,
$z_2 = -2$ and corresponding eigenvectors X_1, X_2

$$C = \begin{pmatrix} 0,2 & -1,8 \\ -2,2 & -0,2 \end{pmatrix}.$$

Then we find expansion into multipliers

$$L(z) = \begin{pmatrix} z-1,8 & 1,2 \\ 2,8 & z-0,2 \end{pmatrix} \begin{pmatrix} z-0,2 & 1,8 \\ 2,2 & z+0,2 \end{pmatrix}.$$

Similarly we find the other expansions into multipliers

$$, \; z_3 = -1, \; L(z) = \begin{pmatrix} z-1,5 & 1,5 \\ 3,5 & z+0,5 \end{pmatrix} \begin{pmatrix} z-0,5 & 1,5 \\ 1,5 & z-0,5 \end{pmatrix};$$

$$z_2 = -2, \; z_4 = 3,$$

$$L(z) = \begin{pmatrix} z-1,25 & 0,75 \\ 2,25 & z+0,25 \end{pmatrix} \begin{pmatrix} z-0,75 & 2,25 \\ 2,75 & z-0,25 \end{pmatrix};$$

$$z_3 = -1, \; z_2 = -2, \; L(z) = \begin{pmatrix} z+1,5 & -1,5 \\ 10,5 & z-6,5 \end{pmatrix} \begin{pmatrix} z-3,5 & 4,5 \\ -5,5 & z+6,5 \end{pmatrix};$$

$$z_3 = -1, \; z_4 = 3, \; L(z) = \begin{pmatrix} z-1 & 1 \\ 3 & z+1 \end{pmatrix} \begin{pmatrix} z-1 & 2 \\ 2 & z-1 \end{pmatrix}.$$

In the general case the polynomial matrix

$$L(z) = Ez^m + \sum_{k=1}^{m} A_k z^{m-k}$$

can sometimes be divided into linear multipliers

$$L(z) = (Ez - C_1)(Ez - C_2)\ldots(Ez - C_m).$$

We shall demonstrate that the solution of the matrix square equation is reduced to expansion of the polynomial matrix

$$L(z) = Ez^2 + A_1 z + A_2 = (Ez - B)(Ez - C).$$

(196)

If expansion (196) is valid, we receive the following system of equations

$$B + C = -A_1,\ BC = A_2,$$

which are reduced to square equations for matrixes B, C

$$B^2 + BA_1 + A_2 = 0,$$

$$C^2 + A_1 C + A_2 = 0.$$

(197)

Example. We solve matrix equations (197) for

$$A_1 = \begin{pmatrix} -3 & 0 \\ 2 & 1 \end{pmatrix}, \qquad A_2 = \begin{pmatrix} 1 & -1 \\ -9 & -9 \end{pmatrix}$$

Using the expansion of polynomial matrix

$$L(z) = \begin{pmatrix} z^2 - 3z + 1 & -1 \\ 2z - z & z^2 - z - 5 \end{pmatrix}$$

we find particular solutions B, C for square equations (197)

$$B = \begin{pmatrix} 1 & -1 \\ -3 & -3 \end{pmatrix}, \quad C = \begin{pmatrix} 2 & 1 \\ 1 & 2 \end{pmatrix}.$$

Note. The question of solution of the square matrix equation is quite complicated. We shall consider, for example, the following square equation

$$X^2 - E = 0,$$

where E is a unit matrix. This equation has the solution

$$X = E - 2P,$$

where P is any projector, $P^2 = P$. In fact

$$X^2 = (E - 2P)^2 = E - 4P + 4P^2 = E - 4P + 4P = E.$$

Thus, the operation of finding a square root of a matrix has an unlimited number of solutions.

In conclusion we shall consider a computational algorithm for finding the spectrum of polynomial matrix [1]

$$A(z) = \sum_{k=0}^{N} z^k A_k, \tag{198}$$

defined by algebraic equation

$$\det A(z) = 0, \tag{199}$$

where A_k $(k = 0, 1, ..., N)$ are matrixes of $m \times m$ size.

We shall compare matrix $A(z)$ with the system of difference equations

$$\sum_{k=0}^{N} A_k X_{n+k} = 0. \tag{200}$$

Equation (199) is multiplicational for the system of difference equations (200).

We find an integral variety of solutions of the system of equations (200)

$$X_{n+1} = Q_n X_n. \tag{201}$$

Matrixes Q_n $(n = 0, \pm1, \pm2, ...)$ satisfy the matrix difference equation

$$A_0 + A_1 Q_n + A_2 Q_{n+1} Q_n + ... + A_N Q_{N-1+n} Q_{N-2+n} ... Q_n = 0. \tag{202}$$

Let z_j $(j = 1, 2, ..., r)$ denote the roots of equation (199), and let C_j denote eigenvectors of the matrix bunch (198), which are found by the equations

$$A(z_j) C_j = 0 \quad (j = 1, ..., r). \tag{203}$$

Let equation (199) have a group of roots $z_1, ..., z_m$ which are the biggest to the modulus and

$$\min_{1 \le j \le m} |z_j| > \max_{m+1 \le j \le r} |z_j|.$$

If the corresponding vectors $C_1, ..., C_s$ are linearly independent then almost any solution of equation (201) has limit Q_+ for $n \to +\infty$ and the corresponding eigenvalues of matrix Q_+ coincide with values z_j $(j = 1,...,m)$.

Let equation (199) have a group of roots $z_{r-s+1},...,z_r$ which are the smallest by the modulus and

$$\min_{1 \leq j \leq r-m} |z_j| > \max_{r-m+1 < j \leq r} |z_j|.$$

If the corresponding eigenvectors $C_{r-m+1},...,C_2$ are linearly independent, then almost any solution of the difference matrix equation (201) has limit Q_- for $n \to -\infty$ and the corresponding eigenvalues of matrix Q_- coincide with values z_j $(j = r-m+1,...,r)$.

The shown results can be applied for finding the roots of equation (199). Thus, the given computational algorithm operates effectively if the biggest by the modulus eigenvalues $z_1,...,z_m$ of matrix bunch (198) (or the smallest by the modulus eigenvalues $z_1,...,z_m$ of matrix bunch (198)) correspond to the linearly independent eigenvectors.

Note. Let us know some solution of the matrix algebraic equation

$$\sum_{k=0}^{N} A_k Q^k = 0,$$

Which comes from equation (202) for $Q_n \equiv Q$ $(n = 0, \pm 1, \pm 2,...)$. At that matrix $A(z)$ can be expanded into multipliers and consequently decrease the order of equation (199). In fact, we have expansion into multipliers

$$A(z) = \sum_{k=0}^{N} A_k E z^k - \sum_{k=0}^{N} A_k Q^k = \left(A_1 + \sum_{k=2}^{N} A_k \left(E z^{k-1} + z^{k-2} Q + ... + Q^{k-1} \right) \right) (Ez - Q),$$

which implies the following equality

$$\det A(z) = \det \left(A_1 + \sum_{k=2}^{N} A_k \left(E z^{k-1} + z^{k-2} Q + ... + Q^{k-1} \right) \right) \det (Ez - Q).$$

Example. We shall consider equation

$$f(z) \equiv \begin{vmatrix} z^2 + 6z & 14,5z^2 + 4z - 1 \\ 1 & z^2 + 4,1z + 1 \end{vmatrix} = 0.$$

We shall create the corresponding system of difference equations

$$AX_{n+2} + BX_{n+1} + CX_n = 0,$$

(204)

where it is denoted

$$A = \begin{pmatrix} 1 & 14,5 \\ 0 & 1 \end{pmatrix}, \quad B = \begin{pmatrix} 0 & 4 \\ 0 & 4,1 \end{pmatrix}, \quad C = \begin{pmatrix} 0 & -1 \\ 1 & 1 \end{pmatrix}.$$

We look for an integral variety of solutions of the system of difference equations (20), found by the system of difference equations

$$X_{n+1} = Q_n X_n.$$

At that we come to the matrix difference equation which can be written like

$$Q_n = -\left(A Q_{n+1} + B \right)^{-1} C; \qquad Q_{n+1} = -A^{-1} \left(B + C Q_n^{-1} \right).$$

For almost any initial values Q_0 there are limits

$$Q_\pm = \lim_{n \to \pm\infty} Q_n,$$

where it is denoted

$$Q_+ = \begin{pmatrix} -8,4745762 & 17,271189 \\ 0,1694915 & -1,5254239 \end{pmatrix},$$

$$Q_- = \begin{pmatrix} 0,125 & 0,5125 \\ -0,250 & -0,2250 \end{pmatrix}.$$

Polynomial $f(z)$ is expanded into two polynomials

$$f(z) = \det\left(Ez - Q_+\right)\det\left(Ez - Q_-\right),$$

which brings to the expansion

$$z^4 + 10,1z^3 + 11,1z^2 + 2z + 1 - \left(z^2 + 10z + 10\right)\left(z^2 + 0,1z + 0,1\right).$$

The numeric algorithms of expansion of random polynomial matrixes into multipliers are also of a great interest.

Questions for self-study

1. Consider functions from several commutative matrixes.

2. Develop and test different numeric methods of defining the rank of the matrix.

3. Develop the asymptotic method in the case when matrix A in the equation system (119) has zero eigenvalues.

4. Consider the asymptotic method in the case of multiple eigenvalues of matrix A in the system of equations (119).

5. Apply the asymptotic method to the system of linear differential equations with coefficients like $\sum_k{}' a_k e^{-\alpha_k t}$, $\operatorname{Re}\alpha_k > 0$.

Literature

1. Valeev K.G. Expansion of the Matrix Spectrum. – Kiev: Vyshcha Shkola, 1986. – 272 p.
2. Gantmacher F.R. Matrix Theory. – M.: Nauka, 1967. – 575 p.
3. Valeev K.G., Finin G.S. Building Lyapunov Functions. – Kiev: Nauk. Dumka, 1981. – 412 p.
4. Valeev K.G., Kurbanshoev S.Z. Building of Integral Varieties. – Dushanbe, Donish, 2006. – 512 p.
5. Stryzhak T.G. Asymptotic Method of Normalization. – Kiev: Vyshcha Shkola, 1984. – 280 p.
6. Barkgoff J.D. Dynamic Systems. – M.: L.: ONTI, 1941. – 320 p.
7. Stryzhak T.G. Approximation Method in the Problems of Mechanics. – Kiev, Donetsk, 1982. – 252 p.

Ассоциация Украинских стипендиатов DAAD
(AUS DAAD)
Национальный комитет IAESTE Украины
(NC IAESTE-Ukraine)

Серия: "Современная математика для инженеров"

Аналитические функции от матрицы

Профессор НТУУ "КПИ"
Тамара Стрижак

(Серия лекций по современным разделам Математики для иностранных стажеров IAESTE)

Проект «Современная математика для инженеров» основан "AUS – DAAD" и Национальным Комитетом IAESTE Украины.

Цель проекта – опубликовать наиболее существенные математические результаты, изложить их в доступном виде для применения инженером, перевести эти работы на английский язык и, прежде всего, через студентов-стажеров IAESTE осуществлять *научный обмен* в области прикладной математики, теории колебаний, теоретической механики т.д.

Импульсом для реализации Проекта «Современная математика для инженеров» послужило то обстоятельство, что 2008 год обьявлен «Годом математики» в Германии.

Начать «Проект» нас вдохновил положительный опыт *ученых Калифорнийского университета*, которые более полвека назад, в 1956 году, опубликовали монографию *«Современная математика для инженеров»**.

Работа имела колоссальный успех. На самом деле эта монография заложила прочный фундамент, на котором надежно и успешно развивается *прикладная математика.*

* - **«Современная математика для инженеров»** под редакцией Э.Ф. Беккенбаха
Перевод с английского под общей редакцией И.Н. Векуа
Издательство иностранная литература, Москва 1958

Функции от матрицы

В работе достаточно полно излагаются функции от квадратной матрицы. Приведены общие сведения о нахождении и свойствах функций от матрицы. Обсуждается проблема собственных чисел матрицы, приближенные способы отыскания собственных чисел. Рассмотрены некоторые вопросы, связанные с теорией матриц.

1. Общие сведения о векторах и матрицах

Поскольку векторы и матрицы изучаются во всех вузах, то приведем лишь те сведения, которые потребуются для понимания этой работы.

Определение. Упорядоченная совокупность m комплексных чисел $x_1, x_2, ..., x_m$ называется m-мерным вектором и обозначается

$$X = \begin{pmatrix} x_1 \\ x_2 \\ ... \\ x_m \end{pmatrix}, \quad X^* = (x_1, x_2, ..., x_m).$$

Вектор X называется вектор-столбцом, X^* называется вектор-строкой. Переход от записи X к записи X^* называется транспонированием вектора. При этом

$$\left(X^*\right)^* = X.$$

Определение. Скалярным произведением двух векторов

$$X = \begin{pmatrix} x_1 \\ x_2 \\ ... \\ x_m \end{pmatrix}, \quad Y = \begin{pmatrix} y_1 \\ y_2 \\ ... \\ y_m \end{pmatrix},$$

называется число

$$Y^*X = X^*Y = (X, Y) = y_1 x_1 + y_2 x_2 + ... + y_m x_m.$$

$$(1)$$

Векторы X, Y называются взаимно ортогональными, если $Y^*X = 0$.

Чтобы характеризировать вектор X одним числом вводят норму $\|X\|$ вектора.

Норма $\|X\|$ – действительное число. Норму вектора можно определить разными способами. При этом должны выполняться следующие свойства

1. $\|X\| \geq 0$. Если $\|X\| = 0$, то $X = 0$,

2. $\|\lambda X\| = |\lambda| \cdot \|X\|$,

3. $\|X + Y\| \leq \|X\| + \|Y\|$.

$$(2)$$

Норму вектора часто определяют следующим образом.

1) $\|X\| = \max_i |x_i| \quad (i = 1, 2, ..., m),$

2) $\|X\|_1 = |x_1| + |x_2| + ... + |x_m| = \sum_{i=1}^{m} |x_i|,$

3) $\|X\|_2 = \sqrt{X^* X} = \sqrt{|x_1|^2 + |x_2|^2 + ... + |x_m|^2}.$

(3)

Все нормы вектора эквивалентны, т.е. при любом определении норм $\|X\|_\alpha$, $\|X\|_\beta$ вектора существуют постоянные c_1, c_2 такие, что

$$c_1 \|X\|_\alpha \leq \|X\|_\beta \leq c_2 \|X\|_\alpha \quad (0 < c_1 \leq c_2).$$

Рассмотрим совокупность векторов $X_1, X_2, ..., X_n$. Эти векторы называются линейно независимыми, если равенство

$$\alpha_1 X_1 + \alpha_2 X_2 + ... + \alpha_n X_n = 0 \qquad (4)$$

выполняется только при $\alpha_1 = 0, \alpha_2 = 0, ..., \alpha_n = 0$.

Пусть $n \leq m$ и векторы $X_1, X_2, ..., X_n$ представлены через проекции

$$X_1 = \begin{pmatrix} x_{11} \\ x_{21} \\ x_{31} \\ .... \\ x_{m1} \end{pmatrix}, \quad X_2 = \begin{pmatrix} 0 \\ x_{22} \\ x_{32} \\ .... \\ x_{m2} \end{pmatrix}, \quad X_3 = \begin{pmatrix} 0 \\ 0 \\ x_{33} \\ .... \\ x_{m3} \end{pmatrix}, \quad ..., \quad X_n = \begin{pmatrix} 0 \\ 0 \\ ... \\ x_{nn} \\ .... \\ x_{mn} \end{pmatrix}.$$

$$(5)$$

Если $x_{11} \neq 0, x_{22} \neq 0, x_{33} \neq 0, ..., x_{nn} \neq 0$, то векторы $X_1, X_2, ..., X_n$ – линейно независимы.

Определение. Система m линейно независимых векторов в m-мерном пространстве называется базисом.

Теорема. Для того, чтобы m векторов

$$X_1 = \begin{pmatrix} x_{11} \\ x_{21} \\ ... \\ x_{m1} \end{pmatrix}, \quad X_2 = \begin{pmatrix} x_{12} \\ x_{22} \\ ... \\ x_{m2} \end{pmatrix}, \quad ..., \quad X_m = \begin{pmatrix} x_{1m} \\ x_{2m} \\ ... \\ x_{mm} \end{pmatrix} \qquad (6)$$

образовывали базис необходимо и достаточно, чтобы определитель Δ, элементами которого являются проекции векторов $X_1, X_2, ..., X_m$,не обращался в нуль:

$$\Delta \equiv \begin{vmatrix} x_{11} & x_{12} & ... & x_{1m} \\ x_{21} & x_{22} & ... & x_{2m} \\ ... & ... & ... & ... \\ x_{m1} & x_{m2} & ... & x_{mm} \end{vmatrix} \neq 0 . \tag{7}$$

Определение. Вектор X_0 называется линейной комбинацией векторов $X_1, X_2, ..., X_m$, если существуют такие числа $\alpha_1, \alpha_2, ..., \alpha_m$, при которых выполняется равенство

$$X_0 = \alpha_1 X_1 + \alpha_2 X_2 + ... + \alpha_m X_m . \tag{8}$$

Вектор $\alpha_k X_k$ называется проекцией вектора X_0 на вектор X_k.

Определение. Пусть V – непустое множество векторов из R^m. Множество V называется подпространством в R^m, если из условия $X \in V$, $Y \in V$ следует, что $\alpha X + \beta Y \in V$, где α, β – комплексные числа.

Определение. Подпространство V, образованное векторами вида

$$X = \alpha_1 X_1 + \alpha_2 X_2 + ... + \alpha_s X_s , \quad \left(s \leq m \right)$$

называется подпространством, порожденным векторами $X_1, X_2, ..., X_s$. Наибольшее число r линейно независимых векторов в системе векторов $X_1, X_2, ..., X_s$ называется ее рангом и обозначается

$$r = rank(X_1, X_2, ..., X_s), \text{ или } r = rang(X_1, X_2, ..., X_s).$$

Вычисление ранга системы векторов является сложной вычислительной задачей при больших значениях m. Обычно систему векторов приводят к виду (5) или используют ортогонализацию векторов.

Определение. Эквивалентными преобразования системы векторов называются следующие преобразования:

1) перестановка векторов;

2) умножение векторов на числа, отличные от нуля;

3) сложение векторов, умноженных на произвольные числа.

Теорема. При эквивалентных преобразованиях системы векторов ее ранг не меняется.

Для вычисления ранга системы векторов используют эквивалентные преобразования для приведения системы векторов к виду, когда ранг легко вычисляется.

Приведем некоторые сведения из теории матриц.

Определение. Матрицей размера $m \times n$ называется прямоугольная таблица чисел

$$A = \begin{pmatrix} a_{11} & a_{12} & \dots & a_{1n} \\ a_{21} & a_{22} & \dots & a_{2n} \\ \dots & \dots & \dots & \dots \\ a_{m1} & a_{m2} & \dots & a_{mn} \end{pmatrix}.$$

Записывают $\dim A = m \times n$. Вектор-строка

$$b = \begin{pmatrix} a_{11} & a_{12} & \dots & a_{1n} \end{pmatrix}$$

является матрицей размера $1 \times n$. Вектор-столбец

$$a = \begin{pmatrix} a_{11} \\ a_{21} \\ \dots \\ a_{m1} \end{pmatrix}$$

является матрицей размера $m \times 1$.

Горизонтальные ряды чисел называются строками матрицы, вертикальные ряды чисел называются столбцами.

Если число строк матрицы равно числу столбцов, то матрица называется квадратной. При этом число строк называется порядком матрицы.

Ряд чисел $a_{11}, a_{22}, \dots, a_{mm}$ называется главной диагональю матрицы.

Определение. Квадратная матрица называется диагональной, если все элементы, расположенные вне главной диагонали, равны нулю.

Диагональная матрица называется единичной, если все элементы, расположенные на главной диагонали, равны единице. Единичная матрица обозначается символом E

$$E = \begin{pmatrix} 1 & 0 & ... & 0 \\ 0 & 1 & ... & 0 \\ ... & ... & ... & ... \\ 0 & 0 & ... & 1 \end{pmatrix}.$$

Определение. Транспонированием матрицы называется замена строк матрицы на столбцы с сохранением порядка их следования. Транспонирование матрицы обозначаем звездочкой *

$$A = \begin{pmatrix} a_{11} & a_{12} & ... & a_{1n} \\ a_{21} & a_{22} & ... & a_{2n} \\ ... & ... & ... & ... \\ a_{m1} & a_{m2} & ... & a_{mn} \end{pmatrix}, \quad A^* = \begin{pmatrix} a_{11} & a_{21} & ... & a_{m1} \\ a_{12} & a_{22} & ... & a_{m2} \\ ... & ... & ... & ... \\ a_{1n} & a_{2n} & ... & a_{mn} \end{pmatrix}.$$

Рассмотрим две матрицы: матрицу A размера $m \times n$ и матрицу B размера $n \times l$

$$A = \begin{pmatrix} a_{11} & a_{12} & ... & a_{1n} \\ a_{21} & a_{22} & ... & a_{2n} \\ ... & ... & ... & ... \\ a_{m1} & a_{m2} & ... & a_{mn} \end{pmatrix}, \quad B = \begin{pmatrix} b_{11} & b_{12} & ... & b_{1l} \\ b_{21} & b_{22} & ... & b_{2l} \\ ... & ... & ... & ... \\ b_{n1} & b_{n2} & ... & b_{nl} \end{pmatrix}.$$

Определение. Произведением матриц AB называется матрица $C = AB$ размера $m \times l$

$$C = \begin{pmatrix} c_{11} & c_{12} & ... & c_{1l} \\ c_{21} & c_{22} & ... & c_{2l} \\ ... & ... & ... & ... \\ c_{m1} & c_{m2} & ... & c_{ml} \end{pmatrix}$$

элементы которой определяются по формулам

$$c_{ks} = \sum_{i=1}^{n} a_{ki} b_{is} \quad (k = 1, 2, ..., m;\ s = 1, 2, ..., l). \quad (9)$$

Отметим некоторые свойства произведения матриц.

1. $\alpha(AB) = (\alpha A) \cdot B = A \cdot (\alpha B).$

2. $(A + B)C = AC + BC.$

3. $C(A + B) = CA + CB.$

4. $ABC = A \cdot (BC) = (AB) \cdot C.$

5. $(AB)^* = B^* A^*.$

6. $AE = A, \quad EA = A.$

В общем случае $AB \neq BA$.

Определение. Если $A \cdot B = E$, $\dim A = m \times m$, $\dim B = m \times m$, то матрица B называется обратной к матрице A и обозначается $B = A^{-1}$.

Если квадратная матрица A имеет обратную матрицу A^{-1}, то она называется неособенной, $\det A \neq 0$ и саму матрицу можно представить в виде

$$A^{-1} = \frac{1}{\Delta} \begin{pmatrix} A_{11} & A_{21} & \dots & A_{m1} \\ A_{12} & A_{22} & \dots & A_{m2} \\ \dots & \dots & \dots & \dots \\ A_{1m} & A_{2m} & \dots & A_{mm} \end{pmatrix}, \qquad (10)$$

$$\Delta = \det \begin{vmatrix} a_{11} & a_{12} & \dots & a_{1m} \\ a_{21} & a_{22} & \dots & a_{2m} \\ \dots & \dots & \dots & \dots \\ a_{m1} & a_{m2} & \dots & a_{mm} \end{vmatrix}.$$

Здесь Δ – определитель матрицы A, A_{ks} – алгебраические дополнения элементов a_{ks} определителя Δ. Отметим, что

$$AA^{-1} = E, \quad A^{-1}A = E, \quad (AB)^{-1} = B^{-1}A^{-1}.$$

2. Норма матрицы

Пусть A – квадратная матрица порядка m. Определим какую-либо норму $\|X\|$ вектора X

$$X = \begin{pmatrix} x_1 \\ x_2 \\ \dots \\ x_m \end{pmatrix}.$$

Нормой матрицы A называется число

$$\max_{\|X\| \leq 1} \|AX\| = \|A\|. \tag{11}$$

Из определения нормы $\|A\|$ следует неравенство

$$\|AX\| \leq \|A\| \cdot \|X\|, \tag{12}$$

которое справедливо при любом векторе X. Из формул (11), (12) следуют следующие свойства нормы матрицы

1. $\|A\| \geq 0$. Если $\|A\| = 0$, то $A = 0$;

2. $\|\alpha A\| = |\alpha| \cdot \|A\|$;

3. $\|A + B\| \leq \|A\| + \|B\|$;

4. $\|AB\| \leq \|A\| \cdot \|B\|$.

Если a_{ks} $\quad (k, s = 1, 2, \dots, m)$ – произвольный элемент матрицы A, то

$$|a_{ks}| \leq \|A\| \quad (k, s = 1, 2, \dots, m). \tag{14}$$

Таким образом, норма матрицы определяется согласовано с определением нормы вектора.

Если $\|X\| = \max_i |x_i|$, $(i = 1,2,...,m)$, то норма матрицы A

$$A = \begin{pmatrix} a_{11} & a_{12} & ... & a_{1m} \\ a_{21} & a_{22} & ... & a_{2m} \\ ... & ... & ... & ... \\ a_{m1} & a_{m2} & ... & a_{mm} \end{pmatrix}$$

определяется по формуле

$$\|A\| = \max_k \sum_{s=1}^{m} |a_{ks}|.$$

Если $\|X\|_1 = |x_1| + |x_2| + ... + |x_m|$, то

$$\|A\|_1 = \max_s \sum_{k=1}^{m} |a_{ks}|.$$

Если $\|X\|_2 = \sqrt{\sum_{k=1}^{m} |x_k|^2}$, то $\|A\|_2 = \sqrt{\lambda}$, где λ – наибольшее значение квадратичной формы

$$f(x) = \sum_{k=1}^{m} (a_{k1}x_1 + a_{k2}x_2 + ... + a_{km}x_m)^2 \qquad \text{при}$$

$|x_1|^2 + |x_2|^2 + ... + |x_m|^2 = 1$. Если введем симметрическую матрицу

$$B = \|b_{ij}\|_1^m, \quad b_{ij} = \sum_{k=1}^{m} a_{ki}a_{kj} \quad (i,j = 1,...,m),$$

то λ является наибольшим собственным числом матрицы B. Отметим, что $B = A^*A$.

Вычислить эвклидову норму $\|A\|_2$ трудно и в вычислениях ее часто заменяют другой нормой матрицы

$$\|A\|_3 = \sqrt{\sum_{k,s=1}^{m} |a_{ks}|^2}, \quad \|A\|_3 \geq \|A\|_2.$$

Пример. Найдем нормы матрицы

$$A = \begin{pmatrix} 1 & 2 \\ 1 & -2 \end{pmatrix}$$

при разных определениях нормы вектора.

$$\|A\| = \max\left\{|1|+|2|,\ |1|+|-2|\right\} = 3;$$

$$\|A\|_1 = \max\left\{|1|+|1|,\ |2|+|-2|\right\} = 4;$$

$$\|A\|_2 = \max_{x_1^2+x_2^2 \leq 1} \sqrt{(x_1+2x_2)^2 + (x_1-2x_2)^2} = \sqrt{8};$$

$$\|A\|_3 = \sqrt{1^2+2^2+1^2+(-2)^2} = \sqrt{10}.$$

Если элементы матрицы A малы по модулю, то при нахождении нормы матрицы $\|A\|_3$ возможны ошибки из-за погрешностей округления чисел.

В качестве приложения понятия нормы матрицы рассмотрим сходимость матричных рядов.

Пусть $A_0, A_1, A_2, ..., A_k, ...$ — матрицы порядка $m \times m$. Выражение

$$A_0 + A_1 + A_2 + ... + A_k + ... \qquad (15)$$

называется матричным рядом. Матричный ряд равносилен множеству m^2 обычных скалярных рядов. Матричный ряд (15) сходится, если существует предел

$$A = \lim_{N \to \infty} \sum_{k=0}^{N} A_k. \qquad (16)$$

В свою очередь матричный ряд (16) сходится абсолютно, если сходится ряд из норм ряда (16)

$$\|A_0\| + \|A_1\| + \|A_2\| + ... + \|A_k\| + ... \qquad (17)$$

Пример. Рассмотрим матричный ряд

$$S = E + A + A^2 + A^3 + ... + A^k + ... \qquad (18)$$

Этот ряд сходится абсолютно, если сходится скалярный ряд

$$\|E\| + \|A\| + \|A^2\| + \|A^3\| + ... + \|A^k\| +$$
$$+ ... \leq 1 + \|A\| + \|A\|^2 + \|A\|^3 + ... + \|A\|^k + ...$$

Следовательно матричный ряд (18) сходится абсолютно при $\|A\| < 1$. Умножим ряд (18) на матрицу $E - A$ и при этом получим равенство

$$S(E - A) = E + A + A^2 + ... +$$
$$+ A^k - A - A^2 - ... - A^{k+1} - ... = E.$$

Следовательно, справедливо равенство

$$E + A + A^2 + A^3 + ... + A^k + ... = (E - A)^{-1}, \quad \|A\| < 1.$$

(19)

Это равенство можно применить для решения системы линейных алгебраических уравнений

$$X = B + AX, \quad \|A\| < 1,$$
(20)

методом последовательных приближений

$$X_0 = 0, \ X_{n+1} = B + AX_n \ (n = 0,1,2,...); \ X = \lim_{n \to \infty} X_n.$$

(21)

Действительно, из векторного уравнения (20) находим

$$X_0 = 0, \quad X_1 = B, \quad X_2 = B + AB,$$

$$X_3 = B + AB + A^2B, \ ...$$

$$X_n = B + AB + A^2B + ... + A^{n-1}B,$$

$$\tilde{O} = \left(E + A + A^2 + ... + A^k + ...\right)B$$

т.е.

$$X = (E - A)^{-1} B.$$

Метод последовательных приближений (21) удобен тем, что возможные вычислительные ошибки при отыскании X_n не влияют на окончательный результат.

3. Собственные векторы и собственные числа матрицы

Рассмотрим квадратную матрицу A порядка m.

Определение. Ненулевой вектор X называется собственным вектором матрицы A, если существует число λ такое, что

$$AX = \lambda X. \qquad (22)$$

Число λ называется собственным числом матрицы A, которое соответствует собственному вектору X. Систему уравнений (22) можно записать в виде

$$AX = \lambda EX, \quad (\lambda E - A)X = 0, \quad X \neq 0. \qquad (23)$$

Собственный вектор X является решением однородной системы линейных алгебраических уравнений. Для существования ненулевого решения $X \neq 0$ системы уравнений (23) необходимо и достаточно, чтобы определитель системы равнялся нулю, т.е.

$$\det(E\lambda - A) = \lambda^m + b_1\lambda^{m-1} + ... + b_m = 0. \qquad (24)$$

Уравнение (24) называется характеристическим уравнением, а его корни являются собственными числами матрицы A. В общем случае собственные числа матрицы и проекции собственных векторов являются комплексными числами.

Из уравнения (24) следует, что определитель матрицы A *равен произведению собственных чисел матрицы* A

$$\lambda_1\lambda_2...\lambda_m = \det A. \qquad (25)$$

Определение. Множество всех собственных чисел матрицы A называется спектром матрицы A и обозначается SrA.

Пример. Найдем собственные числа и собственные векторы матрицы

$$A = \begin{pmatrix} 1 & 4 \\ 1 & 1 \end{pmatrix}.$$

Запишем характеристическое уравнение

$$\det(\lambda E - A) = \begin{vmatrix} \lambda - 1 & -4 \\ -1 & \lambda - 1 \end{vmatrix} = \lambda^2 - 2\lambda - 3 = 0,$$

которое имеет корни $\lambda_1 = -1$, $\lambda_2 = 3$.

При $\lambda = -1$ система уравнений (23) имеет вид

$$\begin{cases} -2x_1 - 4x_2 = 0, \\ -x_1 - 2x_2 = 0, \end{cases} \qquad X_1 = \begin{pmatrix} 2 \\ -1 \end{pmatrix}.$$

При $\lambda = 3$ получим систему уравнений

$$\begin{cases} 2x_1 - 4x_2 = 0, \\ -x_1 + 2x_2 = 0, \end{cases} \qquad X_2 = \begin{pmatrix} 2 \\ 1 \end{pmatrix}.$$

Проверим справедливость равенства (22)

$$AX_1 = \begin{pmatrix} 1 & 4 \\ 1 & 1 \end{pmatrix} \begin{pmatrix} 2 \\ -1 \end{pmatrix} = \begin{pmatrix} -2 \\ 1 \end{pmatrix} = -1 \cdot X_1,$$

$$AX_2 = \begin{pmatrix} 1 & 4 \\ 1 & 1 \end{pmatrix} \begin{pmatrix} 2 \\ 1 \end{pmatrix} = \begin{pmatrix} 6 \\ 3 \end{pmatrix} = 3 \cdot X_2.$$

Спектр матрицы A составляет множество

$$Sr A = \{-1; 3\}.$$

Приведем некоторые свойства собственных векторов.

Теорема. Если все собственные числа матрицы разные, то все собственные векторы матрицы линейно независимы.

Определение. Квадратные матрицы A, B называются подобными, если существует неособенная матрица T такая, что

$$B = T^{-1}AT. \tag{26}$$

Теорема. Если A, B – подобные матрицы, то они имеют одинаковые собственные числа.

Теорема. Если матрица A порядка m имеет разные собственные числа, то матрица A подобна диагональной матрице

$$\Lambda = \begin{pmatrix} \lambda_1 & 0 & \ldots & 0 \\ 0 & \lambda_2 & \ldots & 0 \\ \ldots & \ldots & \ldots & \ldots \\ 0 & 0 & \ldots & \lambda_m \end{pmatrix}. \tag{27}$$

В качестве матрицы T возьмем матрицу, столбцами которой являются собственные числа матрицы A. При этом получим равенство

$$\Lambda = T^{-1}AT$$

или $T\Lambda = AT$. Это матричное равенство распадается на m векторных равенств

$$\lambda_k X_k = AX_k \quad (k = 1, 2, \ldots, m).$$

Определение. Матрица Λ вида (27) называется канонической формой Жордана для матрицы A.

Определение. Если матрица A подобна диагональной матрице вида (27), то говорят, что матрица A имеет простую структуру. Если матрица A имеет различные собственные числа, то они имеют простую структуру. ***В некоторых случаях матрица A с кратными собственными числами может иметь простую структуру.***

Пример. Для матрицы A запишем характеристическое уравнение

$$A = \begin{pmatrix} 1 & -1 \\ 1 & 3 \end{pmatrix},$$

$$\det(\lambda E - A) = \begin{vmatrix} \lambda - 1 & 1 \\ -1 & \lambda - 3 \end{vmatrix} = \lambda^2 - 4\lambda + 4 = 0.$$

Это уравнение имеет корень $\lambda = 2$ кратности 2. Из уравнения

$$\lambda X = AX, \quad (\lambda E - A)X = 0$$

получим систему уравнений для проекций вектора X

$$\begin{cases} (\lambda - 1)x_1 + x_2 = 0, \\ -x_1 + (\lambda - 3)x_2 = 0. \end{cases}$$

Находим ненулевое решение $x_1 = 1$, $x_2 = -1$. Матрица A имеет лишь один собственный вектор

$$X = \begin{pmatrix} 1 \\ -1 \end{pmatrix}.$$

Полагая

$$T = \begin{pmatrix} -1 & 1 \\ 0 & -1 \end{pmatrix}, \quad T^{-1} = \begin{pmatrix} -1 & 1 \\ 0 & -1 \end{pmatrix},$$

находим

$$T^{-1}AT = \begin{pmatrix} 2 & 0 \\ 1 & 2 \end{pmatrix}.$$

Последняя матрица имеет форму Жордана в случае кратного собственного числа.

Пусть матрица A имеет собственные числа $\lambda_1, \lambda_2, ..., \lambda_m$, которым соответствуют собственные векторы $X_1, X_2, ..., X_m$. Предполагая, что эти векторы линейно независимые найдем обратную матрицу T^{-1}

$$T = \left(X_1 X_2 ... X_m\right), \qquad T^{-1} = \begin{pmatrix} Y_1 \\ Y_2 \\ ... \\ Y_m \end{pmatrix}.$$

Строки матрицы T^{-1} обозначаем через $Y_1, Y_2, ..., Y_m$. При этом справедливы равенства $T^{-1}T = E$, или

$$Y_k X_s = \delta_{ks} = \begin{cases} 1 \;\; npu \;\; k = s \\ 0 \;\; npu \;\; k \neq s \end{cases} \left(k, s = 1, 2, ..., m\right).$$

При этом справедлива теорема.

Теорема. Если матрица A имеет m линейно независимых собственных вектора $X_1, X_2, ..., X_m$, то ее можно представить в виде

$$A = \lambda_1 X_1 Y_1 + \lambda_2 X_2 Y_2 + ... + \lambda_m X_m Y_m.$$

Для доказательства достаточно умножить справа матрицу A на вектор X_k

$$A X_k = \sum_{s=1}^{m} \lambda_s X_s \left(Y_s X_k\right) = \lambda_k X_k,$$

что доказывает справедливость теоремы.

Определение. Спектральным радиусом $\rho(A)$ матрицы A называется наибольшее по модулю ее собственное число

$$\rho(A) = \max_j \left\{ \left| \lambda_j \right| \right\}. \tag{28}$$

Спектральный радиус можно определить по формуле

$$\rho(A) = \lim_{k \to +\infty} \sqrt[k]{\left\| A^k \right\|}. \tag{29}$$

Для отыскания спектрального радиуса по формуле (29) можно использовать следующую вычислительную формулу, которую легко реализовать на компьютере.

1. Образуем последовательность матриц и их норм

$$\sigma_1 = \left\| A \right\|, \quad A_1 = A\sigma_1^{-1};$$

$$\sigma_{n+1} = \left\| A_n A_n \right\|, \quad A_{n+1} = A_n A_n \sigma_{n+1}^{-1}.$$

Все матрицы A_k $(k = 1,2,3,...)$ имеют единичную норму.

2. Спектральный радиус $\rho(A)$ находится по формуле

$$\ln \rho(A) = \lim_{N \to +\infty} \left(\ln \sigma_1 + \frac{1}{2}\ln \sigma_2 + \frac{1}{4}\ln \sigma_3 + ... + 2^{-N} \ln \sigma_{N+1} \right).$$

3. Процесс вычислений останавливается, если выполняется неравенство

$$\left| 2^{-N} \ln \sigma_{N+1} \right| < \delta,$$

где δ – заданная точность.

Оценим спектральный радиус. Из равенства (22) находим равенство

$$\|AX\| = \|\lambda X\|,$$

а затем находим неравенство

$$|\lambda| \le \|A\|, \quad \rho(A) \le \|A\|. \tag{30}$$

При любом определении нормы матрицы $\|A\|$, согласованной с нормой вектора, собственные числа матрицы A по модулю меньше $\|A\|$.

4. Проекторы

Для понимания функций от матрицы требуется знание проекторов матрицы. Изложим кратко некоторые сведения о проекторах.

Определение. Матрица P называется идемпотентной или проектором, если выполняется равенство

$$PP = P. \tag{31}$$

В частном случае нулевая и единичная матрицы являются проекторами.

Теорема. Если матрица P является проектором, то матрица $P_1 = E - P$ тоже является проектором.

Доказательство. Найдем произведение

$$P_1 P_1 = (E - P)(E - P) = E - 2P + PP = E - 2P + P = E - P = P_1.$$

Из равенства $P_1 P_1 = P_1$ следует, что матрица P_1 тоже является проектором.

Термин "проектор" можно пояснить следующим образом. Пусть P – матрица линейного преобразования-проектирования вектора на некоторое подпространство. Поскольку повторное проектирование вектора равносильно одноразовому проектированию, то $PP = P$.

Пусть векторы $X_1, X_2, ..., X_m$ образуют базис в пространстве R^m. Разложим произвольный вектор $Y \in R^m$ по базису

$$Y = \alpha_1 X_1 + \alpha_2 X_2 + ... + \alpha_m X_m. \tag{32}$$

Для определения коэффициентов $\alpha_1, \alpha_2, ..., \alpha_m$ получим систему линейных алгебраических уравнений

$$
\begin{aligned}
y_1 &= \alpha_1 x_{11} + \alpha_2 x_{12} + ... + \alpha_m x_{1m}, \\
y_2 &= \alpha_1 x_{21} + \alpha_2 x_{22} + ... + \alpha_m x_{2m}, \\
&.., \\
y_m &= \alpha_1 x_{m1} + \alpha_2 x_{m2} + ... + \alpha_m x_{mm},
\end{aligned}
\qquad
X_k \equiv
\begin{pmatrix}
x_{1k} \\
x_{2k} \\
... \\
x_{mk}
\end{pmatrix},
$$

которую можно записать в векторном виде

$$
Y = TA; \quad T =
\begin{pmatrix}
x_{11} & x_{12} & ... & x_{1m} \\
x_{21} & x_{22} & ... & x_{2m} \\
... & ... & ... & ... \\
x_{m1} & x_{m2} & ... & x_{mm}
\end{pmatrix},
\quad
A =
\begin{pmatrix}
\alpha_1 \\
\alpha_2 \\
... \\
\alpha_m
\end{pmatrix}.
$$

Используя обратную матрицу T, находим вектор A

$$T^{-1}Y = A. \tag{33}$$

Строки матрицы T^{-1} обозначим через $Z_1, Z_2, ..., Z_m$

$$T^{-1} = \begin{pmatrix} Z_1 \\ Z_2 \\ ... \\ Z_m \end{pmatrix}.$$

Поскольку $T^{-1}T = E$, то находим равенство

$$Z_k X_s = \delta_{ks} = \begin{cases} 1, \ \textit{если} \ k = s, \\ 0, \ \textit{если} \ k \neq s; \quad k,s = 1,2,...,m. \end{cases} \tag{34}$$

Из формулы (33) находим коэффициенты α_k

$$\alpha_k = Z_k Y \quad (k = 1,2,...,m).$$

Окончательно находим разложение (32) по векторам базиса

$$Y = X_1 Z_1 \cdot Y + X_2 Z_2 Y + ... + X_m Z_m Y. \tag{35}$$

Матрицы

$$P_k = X_k Z_k \quad (k = 1,2,...,m)$$

являются проекторами в рассматриваемом базисе.

Разложение (32) можно записать в виде

$$Y = P_1 Y + P_2 Y + ... + P_m Y, \tag{36}$$

откуда следует равенство

$$P_1 + P_2 + ... + P_m = E. \tag{37}$$

Найдем произведение разных проекторов

$$P_k P_s = (X_k Z_k)(X_s Z_s) = X_k (Z_k X_s) Z_s = X_k \delta_{ks} Z_s$$
$$(k,s = 1,2,...,m).$$

Окончательно находим равенства

$$P_k P_k = P_k, \quad P_k P_s = 0 \quad (k \neq s;\ k,s = 1,2,...,m). \quad (38)$$

Пример. Пусть в пространстве R^2 выбран базис

$$X_1 = \begin{pmatrix} 1 \\ 2 \end{pmatrix}, \quad X_2 = \begin{pmatrix} 1 \\ 3 \end{pmatrix}.$$

Находим матрицы

$$T = \begin{pmatrix} 1 & 1 \\ 2 & 3 \end{pmatrix}, \quad T^{-1} = \begin{pmatrix} 3 & -1 \\ -2 & 1 \end{pmatrix}$$

и проекторы проектирования на векторы базиса

$$P_1 = \begin{pmatrix} 1 \\ 2 \end{pmatrix}(3 \quad -1) = \begin{pmatrix} 3 & -1 \\ 6 & -2 \end{pmatrix}, \quad P_2 = \begin{pmatrix} 1 \\ 3 \end{pmatrix}(-2 \quad 1) = \begin{pmatrix} -2 & 1 \\ -6 & 3 \end{pmatrix}.$$

Легко проверяется выполнение свойств (36), (38)

$$P_1 P_1 = P_1, \quad P_2 P_2 = P_2, \quad P_1 P_2 = 0, \quad P_2 P_1 = 0, \quad P_1 + P_2 = E.$$

Разложим, например, вектор $Y = \begin{pmatrix} 1 \\ 1 \end{pmatrix}$ по базису

$$Y = P_1 Y + P_2 Y, \quad \begin{pmatrix} 1 \\ 1 \end{pmatrix} = \begin{pmatrix} 3 & -1 \\ 6 & -2 \end{pmatrix}\begin{pmatrix} 1 \\ 1 \end{pmatrix} + \begin{pmatrix} -2 & 1 \\ -6 & 3 \end{pmatrix}\begin{pmatrix} 1 \\ 1 \end{pmatrix}.$$

Имеем разложение

$$\begin{pmatrix} 1 \\ 1 \end{pmatrix} = 2\begin{pmatrix} 1 \\ 2 \end{pmatrix} - 1 \cdot \begin{pmatrix} 1 \\ 3 \end{pmatrix}.$$

Пример. Пусть базис в R^3 образуют векторы

$$X_1 = \begin{pmatrix} 1 \\ 1 \\ 2 \end{pmatrix}, \quad X_2 = \begin{pmatrix} 2 \\ 3 \\ 5 \end{pmatrix}, \quad X_3 = \begin{pmatrix} 3 \\ 4 \\ 6 \end{pmatrix}.$$

Образуем матрицы T, T^{-1}

$$\mathrm{T} = \begin{pmatrix} 1 & 2 & 3 \\ 1 & 3 & 4 \\ 2 & 5 & 6 \end{pmatrix}, \quad \mathrm{T}^{-1} = \begin{pmatrix} 2 & -3 & 1 \\ -1 & 0 & 1 \\ 1 & 1 & -1 \end{pmatrix}.$$

Находим проекторы для разложения по базису

$$P_1 = \begin{pmatrix} 1 \\ 1 \\ 2 \end{pmatrix} \begin{pmatrix} 2 & -3 & 1 \end{pmatrix} = \begin{pmatrix} 2 & -3 & 1 \\ 2 & -3 & 1 \\ 4 & -6 & 2 \end{pmatrix};$$

$$P_2 = \begin{pmatrix} 2 \\ 3 \\ 5 \end{pmatrix} \begin{pmatrix} -2 & 0 & 1 \end{pmatrix} = \begin{pmatrix} -4 & 0 & 2 \\ -6 & 0 & 3 \\ -10 & 0 & 5 \end{pmatrix};$$

$$P_3 = \begin{pmatrix} 3 \\ 4 \\ 6 \end{pmatrix} \begin{pmatrix} 1 & 1 & -1 \end{pmatrix} = \begin{pmatrix} 3 & 3 & -3 \\ 4 & 4 & -4 \\ 6 & 6 & -6 \end{pmatrix}.$$

Легко проверить свойства проекторов

$$P_k P_s = P_k \delta_{ks} \quad (k, s = 1,2,3), \quad P_1 + P_2 + P_3 = \mathrm{E}.$$

Найдем разложения вектора $Y = \begin{pmatrix} 2 \\ 3 \\ 1 \end{pmatrix}$ по базису

$$P_1 Y = \begin{pmatrix} -4 \\ -4 \\ -8 \end{pmatrix} = -4X_1, \quad P_2 Y = \begin{pmatrix} -6 \\ -9 \\ -15 \end{pmatrix} = -3X_2,$$

$$P_3 Y = \begin{pmatrix} 12 \\ 16 \\ 24 \end{pmatrix} = 4X_3.$$

Отсюда находим разложение

$$Y = -4X_1 - 3X_2 + 4X_3.$$

Если известны проекторы, то разложение вектора по базису сводится к умножению вектора на проекторы.

Замечание. Пусть P – некоторый проектор, т.е. $PP = P$. Тогда матрица

$$B = 2P - \mathrm{E}$$

является корнем из единичной матрицы, т.е.

$$B = \sqrt{\mathrm{E}}, \quad B^2 = \mathrm{E}.$$

Действительно имеем равенство

$$B \cdot B = (2P - \mathrm{E})(2P - \mathrm{E}) = 4P^2 - 4P + \mathrm{E} = \mathrm{E}.$$

Обратно, если $B = \sqrt{\mathrm{E}}$ – любой корень из единичной матрицы, то матрица

$$P = \frac{1}{2}\left(\mathrm{E} + \sqrt{\mathrm{E}}\right)$$

является проектором. Действительно, имеем равенство

$$PP = \frac{1}{4}\left(E + 2E\sqrt{E} + E\right) = \frac{1}{2}\left(E + \sqrt{E}\right) = P.$$

Пример. Матрица

$$P = \begin{pmatrix} 3 & -1 \\ 6 & -2 \end{pmatrix}$$

является проектором. При этом матрица

$$B = 2P - E = \begin{pmatrix} 5 & -2 \\ 12 & -5 \end{pmatrix}$$

является корнем из единичной матрицы.

Пусть матрица A имеет линейно независимые собственные векторы $X_1, X_2, ..., X_m$. Если базис образован собственными векторами матрицы A, то соответствующие проекторы называем проекторами матрицы A.

Если матрица A имеет собственные числа $\lambda_1, \lambda_2, ..., \lambda_m$, то можем восстановить саму матрицу по формуле

$$A = \lambda_1 P_1 + \lambda_2 P_2 + ... + \lambda_m P_m. \qquad (39)$$

Действительно, находим произведение

$$AX_k = \left(\lambda_1 P_1 + \lambda_2 P_2 + ... + \lambda_m P_m\right) X_k = \lambda_k X_k,$$

так как $P_s X_k = 0$ $(s \neq k)$. Следовательно, X_k – собственный вектор, а λ_k – собственное число матрицы A.

Пример. Пусть матрица A второго порядка имеет собственные векторы

$$X_1 = \begin{pmatrix} 1 \\ 2 \end{pmatrix}, \quad X_2 = \begin{pmatrix} 1 \\ 3 \end{pmatrix},$$

которые соответствуют собственным числам $\lambda_1 = 1$, $\lambda_2 = 2$.

Находим проекторы матрицы A

$$P_1 = \begin{pmatrix} 3 & -1 \\ 6 & -2 \end{pmatrix} \qquad\qquad P_2 = \begin{pmatrix} -2 & 1 \\ -6 & 3 \end{pmatrix}$$

и саму матрицу A

$$A = \lambda_1 P_1 + \lambda_2 P_2 = 1 \cdot \begin{pmatrix} 3 & -1 \\ 6 & -2 \end{pmatrix} + 2 \cdot \begin{pmatrix} -2 & 1 \\ -6 & 3 \end{pmatrix} = \begin{pmatrix} -1 & 1 \\ -6 & 4 \end{pmatrix}.$$

5. Многочлены от матрицы

Пусть A – квадратная матрица размера $m \times m$. Возьмем произвольный многочлен

$$\varphi(z) = a_0 + a_1 z + a_2 z^2 + \ldots + a_k z^k.$$

Определение. Многочленом $\varphi(A)$ от матрицы A называется матрица

$$\varphi(A) = a_0 E + a_1 A + a_2 A^2 + \ldots + a_k A^k. \qquad (40)$$

Пусть X_s – собственный вектор матрицы A, который соответствует собственному числу λ_s. Тогда справедливы равенства

$$E X_s = X_s, \quad A X_s = \lambda_s X_s, \quad A^2 X_s = \lambda_s^2 X_s, \ldots,$$
$$A^k X_s = \lambda_s^k X_s.$$

Отсюда находим равенство

$$\varphi(A) X_s = \left(a_0 E + a_1 A + a_2 A^2 + \ldots + a_k A^k\right) X_s =$$
$$= a_0 X_s + a_1 \lambda_s X_s + a_2 \lambda_s^2 X_s + \ldots + a_k \lambda_s^k X_s =$$
$$= \left(a_0 + a_1 \lambda_s + a_2 \lambda_s^2 + \ldots + a_k \lambda_s^k\right) X_s = \varphi(\lambda_s) X_s.$$

Отсюда следует справедливость теоремы.

Теорема. Многочлен $\varphi(A)$ есть матрица, которая имеет те же собственные векторы X_s, что и матрица A и соответствующие собственные числа $\varphi(\lambda_s)$.

Пусть $f(\lambda)$ – характеристический многочлен матрицы A

$$f(\lambda) \equiv \det(E\lambda - A).$$

Из предыдущего следует теорема Гамильтона-Кэли.

Теорема. Любая квадратная матрица является корнем своего характеристического многочлена [2].

Доказательство. Пусть матрица A имеет различные собственные числа $\lambda_1, \lambda_2, ..., \lambda_m$, которым соответствуют линейно независимые собственные векторы $X_1, X_2, ..., X_m$. Собственные числа $\lambda_1, \lambda_2, ..., \lambda_m$ являются корнями характеристического многочлена, т.е. $f(\lambda_s) = 0$, $(s = 1, 2, ..., m)$. Имеем равенства

$$f(A)X_s = f(\lambda_s)X_s = 0, \quad (s = 1, 2, ..., m). \quad (41)$$

Введем матрицу T, столбцами которой являются собственные векторы X_s матрицы A

$$T = (X_1 X_2 ... X_m), \quad \det T \neq 0.$$

Из равенств (41) получим матричное равенство

$$f(A)T = 0. \quad (42)$$

Поскольку матрица T невырожденная, то существует обратная матрица T^{-1}. Умножим справа равенство (42) на матрицу T^{-1} и приходим к равенству $f(A) = 0$.

Пример. Проверим правильность теоремы Гамильтона-Кэли на примере матрицы второго порядка

$$A = \begin{pmatrix} 1 & 2 \\ 3 & 4 \end{pmatrix}, \quad \det(E\lambda - A) \equiv \begin{vmatrix} \lambda - 1 & -2 \\ -3 & \lambda - 4 \end{vmatrix} = \lambda^2 - 5\lambda - 2.$$

Найдем значения матричного многочлена от матрицы

$$f(A) \equiv A^2 - 5A - 2E =$$

$$= \begin{pmatrix} 7 & 10 \\ 15 & 12 \end{pmatrix} - 5\begin{pmatrix} 1 & 2 \\ 3 & 4 \end{pmatrix} - 2\begin{pmatrix} 1 & 0 \\ 0 & 1 \end{pmatrix} = \begin{pmatrix} 0 & 0 \\ 0 & 0 \end{pmatrix}.$$

В результате получили $f(A) = 0$.

Рассмотрим теперь аналитическую функцию $\varphi(z)$, представимую рядом Тейлора

$$\varphi(z) = a_0 + a_1 z + a_2 z^2 + \ldots + a_k z^k + \ldots, \tag{43}$$

который сходится в круге $|z| < R$, содержащем все собственные числа $\lambda_1, \lambda_2, \ldots, \lambda_m$ матрицы A. Аналитическая функция $\varphi(A)$ от матрицы A определяется по формуле

$$\varphi(A) = a_0 E + a_1 A + a_2 A^2 + \ldots + a_k A^k + \ldots. \tag{44}$$

Из свойств проекторов P_k матрицы A и формулы (39) вытекает формула для степеней матрицы A

$$A^n = \lambda_1^k P_1 + \lambda_2^k P_2 + \ldots + \lambda_m^k P_m. \tag{45}$$

Ряд (44) можно с помощью формулы (45) записать в виде

$$\varphi\left(\mathrm{A}\right)=\sum_{k=0}^{\infty}a_{k}\left(\lambda_{1}^{k}P_{1}+\lambda_{2}^{k}P_{2}+...+\lambda_{m}^{k}P_{m}\right)=$$

$$=\sum_{s=1}^{m}\left(\sum_{k=0}^{\infty}a_{k}\lambda_{s}^{k}\right)P_{s}=\sum_{s=1}^{m}\varphi\left(\lambda_{s}\right)P_{s}$$

Окончательно получим формулу для аналитической функции от матрицы A

$$\varphi(\mathrm{A})=\sum_{s=1}^{m}\varphi(\lambda_{s})P_{s}\,. \qquad (46)$$

Матрица $\varphi(\mathrm{A})$ имеет те же собственные векторы $X_{1},X_{2},...,X_{m}$, что и матрица A и им соответствуют собственные числа $\phi(\lambda_{1}),\phi(\lambda_{2}),...,\phi(\lambda_{m})$.

Представление функции от матрицы (46) называется спектральным расщеплением функции от матрицы A простой структуры, когда число собственных векторов равно порядку матрицы.

Пример. Найдем спектральное расщепление функции от матрицы A

$$\mathrm{A}=\begin{pmatrix}1 & 2\\ 4 & 3\end{pmatrix}.$$

Найдем собственные числа λ_{1}, λ_{2} матрицы A из уравнения

$$\det\left(E\lambda - A\right) \equiv \begin{vmatrix} \lambda - 1 & -2 \\ -4 & \lambda - 3 \end{vmatrix} = \lambda^2 - 4\lambda - 5 = 0 \, ;$$

$$\lambda_1 = 5 \, , \quad \lambda_2 = -1$$

и соответствующие им собственные векторы

$$X_1 = \begin{pmatrix} 1 \\ 2 \end{pmatrix}, \quad X_2 = \begin{pmatrix} -1 \\ 1 \end{pmatrix}.$$

Строим матрицы T, T^{-1} и проекторы P_1, P_2

$$T = \begin{pmatrix} 1 & -1 \\ 2 & 1 \end{pmatrix}, \quad T^{-1} = \frac{1}{3}\begin{pmatrix} 1 & 1 \\ -2 & 1 \end{pmatrix}$$

$$P_1 = \begin{pmatrix} 1 \\ 2 \end{pmatrix}\begin{pmatrix} \dfrac{1}{3} & \dfrac{1}{3} \end{pmatrix} = \frac{1}{3}\begin{pmatrix} 1 & 1 \\ 2 & 2 \end{pmatrix},$$

$$P_2 = \begin{pmatrix} -1 \\ 1 \end{pmatrix}\begin{pmatrix} \dfrac{-2}{3} & \dfrac{1}{3} \end{pmatrix} = \frac{1}{3}\begin{pmatrix} 2 & -1 \\ -2 & 1 \end{pmatrix}.$$

По формуле (46) находим общее выражение для функции от матрицы A

$$\varphi\left(A\right) = \varphi\left(5\right)\frac{1}{3}\begin{pmatrix} 1 & 1 \\ 2 & 2 \end{pmatrix} + \varphi\left(-1\right)\frac{1}{3}\begin{pmatrix} 2 & -1 \\ -2 & 1 \end{pmatrix}.$$

В частном случае для функции $\varphi\left(z\right) = z^n$ получим равенство

$$A^n = 5^n\,\frac{1}{3}\begin{pmatrix} 1 & 1 \\ 2 & 2 \end{pmatrix} + \left(-1\right)^n\frac{1}{3}\begin{pmatrix} 2 & -1 \\ -2 & 1 \end{pmatrix}.$$

Для функции $\varphi\left(z\right) = e^{zt}$ получим формулу

$$e^{At} = e^{5t}\ \frac{1}{3}\begin{pmatrix} 1 & 1 \\ 2 & 2 \end{pmatrix} + e^{-t}\ \frac{1}{3}\begin{pmatrix} 2 & -1 \\ -2 & 1 \end{pmatrix}.$$

Найдем обратную матрицу A^{-1} с помощью функции

$$\varphi(z) = \frac{1}{z}$$

$$A^{-1} = \frac{1}{15}\begin{pmatrix} 1 & 1 \\ 2 & 2 \end{pmatrix} - \frac{1}{3}\begin{pmatrix} 2 & -1 \\ -2 & 1 \end{pmatrix} = \frac{1}{5}\begin{pmatrix} -3 & 2 \\ 4 & -1 \end{pmatrix}.$$

6. Формула Лагранжа для функции от матрицы

Пусть матрица A имеет простую структуру и число линейно независимых собственных векторов равно порядку матрицы. Из формулы (46) следует, что значения двух аналитических функций $\varphi(A)$, $\psi(A)$ будут совпадать, если значения функций $\varphi(z)$, $\psi(z)$ совпадают на спектре матрицы A, т.е.

$$\varphi(\lambda_k) = \psi(\lambda_k) \qquad (k = 1,2,...,m).$$

При этом $\varphi(A) \equiv \psi(A)$.

При заданной функции $\varphi(A)$ от матрицы A построим интерполяционный многочлен $g(z)$, который в точках λ_k принимает значения $\varphi(\lambda_k)$, $(k = 1,2,...,m)$

$$g(z) = \frac{(z-\lambda_2)\ (z-\lambda_3)...(z-\lambda_m)}{(\lambda_1-\lambda_2)\ (\lambda_1-\lambda_3)...(\lambda_1-\lambda_m)}\ \varphi(\lambda_1) +$$

$$+\frac{(z-\lambda_1)\ (z-\lambda_3)...(z-\lambda_m)}{(\lambda_2-\lambda_1)\ (\lambda_2-\lambda_3)...(\lambda_2-\lambda_m)}\ \varphi(\lambda_2) +...+$$

$$+\frac{(z-\lambda_1)\ (z-\lambda_2)...(z-\lambda_{m-1})}{(\lambda_m-\lambda_1)\ (\lambda_m-\lambda_2)...(\lambda_m-\lambda_{m-1})}\ \varphi(\lambda_m).$$

Подставим вместо z матрицу A и получим формулу для функции $\varphi(A)$ от матрицы A

$$\varphi(A) = \frac{(A - \lambda_2 E)\ (A - \lambda_3 E)...(A - \lambda_m E)}{(\lambda_1 - \lambda_2)\ (\lambda_1 - \lambda_3)...(\lambda_1 - \lambda_m)}\ \varphi(\lambda_1) +$$

$$+\frac{(A - \lambda_1 E)\ (A - \lambda_3 E)...(A - \lambda_m E)}{(\lambda_2 - \lambda_1)\ (\lambda_2 - \lambda_3)...(\lambda_2 - \lambda_m)}\ \varphi(\lambda_2) + ... + \tag{47}$$

$$+\frac{(A - \lambda_1 E)\ (A - \lambda_2 E)...(A - \lambda_{m-1} E)}{(\lambda_m - \lambda_1)\ (\lambda_m - \lambda_2)...(\lambda_m - \lambda_{m-1})}\ \varphi(\lambda_m).$$

Сравнивая полученную формулу с формулой (46) получим выражение для проекторов P_k $(k = 1,2,...,m)$ через матрицу A и собственные числа λ_k $(k = 1,2,...,m)$ матрицы A

$$P_k = \prod_{s=1,s \neq k}^{m} \frac{A - \lambda_s E}{\lambda_k - \lambda_s} \quad (k = 1,2,...,m). \tag{48}$$

Поэтому все проекторы P_k матрицы A являются многочленами степени $(m-1)$ от матрицы A.

Теорема. Найдем проекторы матрицы

$$A = \begin{pmatrix} 1 & 2 \\ 4 & 3 \end{pmatrix}, \quad \lambda_1 = 5, \quad \lambda_2 = -1.$$

По формуле (48) получим выражения

$$P_1 = \frac{A - \lambda_2 E}{\lambda_1 - \lambda_2} = \frac{1}{6}\begin{pmatrix} 2 & 2 \\ 4 & 4 \end{pmatrix}, \quad P_2 = \frac{A - \lambda_1 E}{\lambda_2 - \lambda_1} = -\frac{1}{6}\begin{pmatrix} -4 & 2 \\ 4 & -2 \end{pmatrix}.$$

Пусть $f(z)$ – характеристический многочлен матрицы A

$$f(z) \equiv (z - \lambda_1)\ (z - \lambda_2)...(z - \lambda_m).$$

Из теоремы Гамильтона-Кэли следует равенство

$$f(A) \equiv (A - \lambda_1 E)\ (A - \lambda_2 E)...(A - \lambda_m E).$$

Из формулы (48) следует, что

$$(A - \lambda_k E)P_k = 0; \quad P_k(A - \lambda_k E) = 0. \qquad (49)$$

Следовательно, все столбцы матрицы P_k пропорциональны правому собственному вектору X_k матрицы A, а все строки проектора P_k пропорциональны левому собственному вектору матрицы A.

Отметим, что ранги проекторов P_k $(k = 1,2,...,m)$ равны единице.

Приведем еще один способ построения проекторов матрицы A. Рассмотрим вспомогательную функцию

$$\varphi(z) = \frac{1}{\lambda - z}. \qquad (50)$$

При достаточно большом значении $|\lambda|$ функция $\varphi(z)$ аналитична в круге, содержащем собственные числа $\lambda_1, \lambda_2, ..., \lambda_m$ матрицы A. По формуле (46) находим равенство

$$(E\lambda - A)^{-1} = \frac{1}{\lambda - \lambda_1}P_1 + \frac{1}{\lambda - \lambda_2}P_2 + ... + \frac{1}{\lambda - \lambda_m}P_m. \ (51)$$

Для отыскания проекторов можно использовать вычеты

$$P_k = res(\text{E}\lambda - \text{A})^{-1} \quad \text{при} \quad \lambda = \lambda_k. \tag{52}$$

Левую часть равенства (51) можно вычислить с помощью присоединенной матрицы $B(\lambda)$, элементы которой $b_{ks}(\lambda)$ являются алгебраическими дополнениями элемента $\delta_{sk}\lambda - a_{sk}$ матрицы $\text{E}\lambda - \text{A}$ $\left(\delta_{kk} = 1; \ \delta_{sk} = 0; \ s \neq k\right)$.

Для обратной матрицы имеем равенство

$$(\text{E}\lambda - \text{A})^{-1} = \frac{B(\lambda)}{f(\lambda)}; \quad f(\lambda) = \det(\text{E}\lambda - \text{A}), \tag{53}$$

где $f(\lambda)$–характеристический многочлен матрицы A. Отыскивая вычеты в точках $\lambda = \lambda_k$, приходим к известным формулам для проекторов матрицы A [2]

$$P_k = \operatorname*{res}_{\lambda = \lambda_k} \frac{B(\lambda)}{f(\lambda)}; \quad P_k = \frac{B(\lambda_k)}{f'(\lambda_k)} \quad (k = 1, 2, \ldots, m);$$

$$f'(\lambda) \equiv \frac{df(\lambda)}{d\lambda}. \tag{54}$$

Пример. Найдем проекторы матрицы

$$\text{A} = \begin{pmatrix} 0 & 2 \\ -1 & 3 \end{pmatrix}.$$

Определяем характеристический многочлен $f(\lambda)$, собственные числа λ_1, λ_2 матрицы A и присоединенную матрицу $B(\lambda)$

$$f(\lambda) = \det(E\lambda - A) \equiv \begin{vmatrix} \lambda & -2 \\ 1 & \lambda-3 \end{vmatrix} \equiv \lambda^2 - 3\lambda + 2\,;$$

$$f(\lambda) = 0\,;\ \lambda_1 = 1,\ \lambda_2 = 2\,;\quad B(\lambda) = \begin{pmatrix} \lambda-3 & 2 \\ -1 & \lambda \end{pmatrix}.$$

По формуле (54) вычисляем проекторы

$$P_1 = \frac{B(1)}{f'(1)} = \begin{pmatrix} 2 & -2 \\ 1 & -1 \end{pmatrix};\quad P_2 = \frac{B(2)}{f'(2)} = \begin{pmatrix} -1 & 2 \\ -1 & 2 \end{pmatrix}.$$

Любая аналитическая функция $\varphi(A)$ от матрицы A находится по формуле (46)

$$\varphi(A) = \varphi(1)\begin{pmatrix} 2 & -2 \\ 1 & -1 \end{pmatrix} + \varphi(2)\begin{pmatrix} -1 & 2 \\ -1 & 2 \end{pmatrix}.$$

В частности при $\varphi(z) = \ln z$ получим

$$\ln A = \ln 2 \cdot \begin{pmatrix} -1 & 2 \\ -1 & 2 \end{pmatrix}.$$

7. Формула Коши для функции от матрицы

Пусть аналитическая функция $\varphi(z)$ аналитична в некоторой односвязной области D, содержащей спектр матрицы A – совокупность всех собственных чисел матрицы A. Определим значение аналитической функции $\varphi(z)$ по формуле

(46), где $\varphi(z)$ – значения функции $\varphi(z)$, аналитически продолженной от некоторой точки области D. Пусть замкнутый без самопересечений контур Γ охватывает весь спектр матрицы A. В этом случае можно определить значения функции $\varphi(\lambda)$ с помощью формулы Коши

$$\varphi(\lambda) = \frac{1}{2\pi i}\int_{\tilde{A}} \varphi(z)\,(z-\lambda)^{-1}\,dz. \qquad (55)$$

Формулу (46) можно записать в виде формулы

$$\varphi(A) = \sum_{k=1}^{m}\frac{1}{2\pi i}\int_{\tilde{A}} \frac{\varphi(z)}{z-\lambda_k}\,dz \cdot P_k,$$

которая с учетом равенства (51) принимает вид

$$\varphi(A) = \frac{1}{2\pi i}\int_{\tilde{A}} \varphi(z)\,(Ez-A)^{-1}\,dz. \qquad (56)$$

Формула (56) аналогична формуле Коши (55). Вместо числа λ поставлена матрица A. Эта формула справедлива для аналитической функции $\varphi(A)$ от матрицы A. Из формулы (56) видно, что аналитическая функция от матрицы является непрерывной функцией от матрицы A, т.е. при непрерывном изменении элементов матрицы A элементы матрицы $\varphi(A)$ изменяются тоже непрерывно.

Из формулы (51) следует, что проекторы матрицы A можно определить по формуле

$$P_k = \frac{1}{2\pi i} \int\limits_{\Gamma_k} \left(\mathbf{E}z - \mathbf{A}\right)^{-1} dz \qquad \left(k = 1,2,...,m\right), \qquad (57)$$

где Γ_k – замкнутый контур, охватывающий лишь точку $z = \lambda_k$ и не содержит внутри себя других точек спектра $z = \lambda_s$ $\left(s \neq k;\ s = 1,2,...,m\right)$.

Если некоторый контур γ охватывает сразу несколько точек спектра $\lambda_1,...,\lambda_s$, то находим сумму соответствующих проекторов по формуле

$$P_1 + ... + P_s = \frac{1}{2\pi i} \int\limits_{\gamma} \left(\mathbf{E}z - \mathbf{A}\right)^{-1} dz .$$

При выводе формулы (56) предполагалось, что матрица $\mathbf{A}$ имеет простые собственные числа. Дадим другой вывод формулы (56), не использующей предположения о простоте структуры матрицы $\mathbf{A}$.

Пусть функция $\varphi(z)$ аналитична в замкнутом круге $|z| \leq R$ и $\|\mathbf{A}\| < R$. При этом все собственные числа матрицы $\mathbf{A}$ лежат внутри круга $|z| \leq R$. Разложим функцию $\varphi(z)$ в ряд Тейлора

$$\varphi(z) = \varphi(0) + \frac{\varphi'(0)}{1!}z + \frac{\varphi''(0)}{2!}z^2 + ... + \frac{\varphi^{(n)}(0)}{n!}z^n + ...,$$

где коэффициенты определяются с помощью формул Коши

$$\frac{\varphi^{(n)}(0)}{n!} = \frac{1}{2\pi i} \int\limits_{|z|=R} \frac{\varphi(z)}{z^{n+1}} dz \quad (n = 0,1,2,\ldots). \quad (58)$$

Разложение функции $\varphi(A)$ от матрицы A в ряд

$$\varphi(A) = \varphi(0)E + \frac{\varphi'(0)}{1!}A + \frac{\varphi''(0)}{2!}A^2 + \ldots + \frac{\varphi^{(n)}(0)}{n!}A^n + \ldots$$

можно записать в виде

$$\varphi(A) = \frac{1}{2\pi i} \int\limits_{|z|=R} \varphi(z)\left(\frac{E}{z} + \frac{A}{z^2} + \frac{A^2}{z^3} + \ldots + \frac{A^n}{z^{n+1}} + \ldots\right)dz.$$

$$(59)$$

В силу предположения о норме матрицы A ряд

$$(Ez - A)^{-1} = \frac{E}{z} + \frac{A}{z^2} + \frac{A^2}{z^3} + \ldots + \frac{A^n}{z^{n+1}} + \ldots$$

равномерно сходится на окружности $|z| = R$ и может быть почленно проинтегрирован. Получим формулу для произвольной аналитической функции от матрицы A

$$\varphi(A) = \frac{1}{2\pi i} \int\limits_{|z|=R} f(z)\, (Ez - A)^{-1} dz. \quad (60)$$

Эта формула является матричным аналогом формулы Коши (55). Контур $|z| = R$ можно заменить другим контуром Γ, содержащим спектр матрицы A. Элементы матрицы $\varphi(A)$ аналитически зависят от элементов матрицы A.

Пусть функция $\varphi(z)$ аналитически продолжена из круга $|z| \leq R$ в некоторую односвязную область D с границей Γ, содержащую все собственные числа $\lambda_1, \lambda_2, ..., \lambda_m$ матрицы A. При этом приходим к формуле (55), где уже не использовалось предположение о простоте структуры матрицы A. С помощью формулы (55) можно получить спектральное расщепление функции от матрицы в общем случае

Пусть матрица A имеет собственные числа $\lambda_1, ..., \lambda_q$ кратности $m_1, .., m_q$, $m_1 + ... + m_q = m$. Характеристический многочлен $f(z)$ разлагается на множители

$$f(z) \equiv \det(Ez - A) = (z - z_1)^{m_1} ... (z - z_q)^{m_q}.$$

Находим аналитическое представление обратной матрицы $(Ez - A)^{-1}$ с помощью разложения элементов матрицы на простейшие дроби

$$(Ez - A)^{-1} = \frac{B(z)}{f(z)} = \sum_{k=1}^{q} \left(\frac{B_{k1}}{z - \lambda_k} + \frac{B_{k2} \cdot 1!}{(z - \lambda_k)^2} + ... + \frac{B_{km_k}(m_k - 1)!}{(z - \lambda_k)^{m_k}} \right)$$

.

По формуле Коши находим основную формулу для функции от произвольной матрицы [2]

$$\phi(A) = \sum_{k=1}^{q} \left(f(\lambda_k) B_{k1} + f'(\lambda_k) B_{k2} + \dots + f^{(m_k-1)}(\lambda_k) B_{km_k} \right), (61)$$

которою называют также спектральным расщеплением функции от матрицы A. Матрицы B_{ks} называют компонентами матрицы A. Они не зависят от вида функции $\varphi(z)$ и определяются полностью матрицей A. Матрицы B_{ks} можно выразить через присоединенную матрицу $B(z)$

$$B_{ks} = \lim_{z \to \lambda_k} \frac{1}{r!(s-1)!} \frac{d^r}{dz^r} \left(\frac{B(z)(z-\lambda_k)^{m_k}}{f(z)} \right)$$

$$\left(k = 1, \dots, q; \ s = 1, \dots, m_k; \ r \equiv m_k - s \right).$$

Матрицы B_{ks} можно найти с помощью интерполяционных формул, когда значения интерполяционного многочлена $g(z)$ совпадают на спектре матрицы A (с учетом кратности собственных чисел) со значениями функции $\varphi(z)$, т.е. выполнены равенства

$$g(\lambda_k) = \varphi(\lambda_k), \quad g'(\lambda_k) = \varphi'(\lambda_k), \ \dots,$$
$$g^{(m_k-1)}(\lambda_k) = \phi^{(m_k-1)}(\lambda_k) \quad \left(k = 1, \dots, q \right).$$

В частных случаях можно подобрать такие простые функции $\varphi(z)$, как правило полиномы, когда легко можно вычислить левую часть в формуле (61), а затем последовательно определить компоненты матрицы A. Отметим, что некоторые из матриц B_{ks} могут оказаться нулевыми.

Пример. Применим формулу (61) для отыскания функции от матрицы

$$A = \begin{pmatrix} -1 & 2 \\ -2 & 3 \end{pmatrix}$$

с кратным собственным числом $\lambda_1 = 1$, ($m_1 = 2$). Для аналитической функции $\varphi(z)$ аналитичной в точке $z = 1$ имеем представление вида (61)

$$\varphi(A) = \varphi(1)B_1 + \varphi'(1)B_2, \qquad (62)$$

где матрицы B_1, B_2 не зависят от вида функции $\varphi(z)$.

Возьмем две функции $\varphi_1(z) \equiv 1$, $\varphi_2(z) \equiv z$. Из формулы (62) получим равенства

$$\begin{cases} E = 1 \cdot B_1 + 0 \cdot B_2 \\ A = 1 \cdot B_1 + 1 \cdot B_2 \end{cases}$$

откуда находим: $B_1 = E$, $B_2 = A - E$. Окончательно находим формулу для функции от матрицы A

$$\varphi(A) = \varphi(1) \cdot E + \varphi'(1)\ (A - E)$$

или

$$\varphi(A) = \varphi(1) \begin{pmatrix} 1 & 0 \\ 0 & 1 \end{pmatrix} + \varphi'(1) \begin{pmatrix} -2 & 2 \\ -2 & 2 \end{pmatrix}.$$

Для проверки вычислим матрицы A^2, A^{-1} непосредственно и по формуле (62). Имеем

$$A^2 = \begin{pmatrix} -1 & 2 \\ -2 & 3 \end{pmatrix} \begin{pmatrix} -1 & 2 \\ -2 & 3 \end{pmatrix} = \begin{pmatrix} -3 & 4 \\ -4 & 5 \end{pmatrix}, \quad A^{-1} = \begin{pmatrix} 3 & -2 \\ 2 & -1 \end{pmatrix}.$$

По формуле (62) находим теже значения

$$A^2 = 1 \cdot \begin{pmatrix} 1 & 0 \\ 0 & 1 \end{pmatrix} + 2 \cdot \begin{pmatrix} -2 & 2 \\ -2 & 2 \end{pmatrix} = \begin{pmatrix} -3 & 4 \\ -4 & 5 \end{pmatrix};$$

$$A^{-1} = 1 \cdot \begin{pmatrix} 1 & 0 \\ 0 & 1 \end{pmatrix} - 1 \cdot \begin{pmatrix} -2 & 2 \\ -2 & 2 \end{pmatrix} = \begin{pmatrix} 3 & -2 \\ 2 & -1 \end{pmatrix}.$$

Замечание. Если в формуле (61) возьмем функцию $\varphi(z) \equiv z$, то получим известное представление матрицы A через ее компоненты

$$A = \sum_{k=1}^{q} (\lambda_k B_{k1} + B_{k2}). \qquad (63)$$

Пример. Рассмотрим матрицу третьего порядка

$$A = \begin{pmatrix} -2 & -3 & 3 \\ -4 & -3 & 4 \\ -6 & -6 & 7 \end{pmatrix}.$$

Характеристический многочлен

$$f(z) = \det(Ez - A) = z^3 - 2z^2 + z = (z-1)^2 z$$

имеет двукратный корень $z_1 = 1$ и однократный корень $z_2 = 0$.

Любая аналитическая функция $\varphi(A)$ от матрицы A имеет вид

$$\varphi(A) = \varphi(1)B_{11} + \varphi'(1)B_{12} + \varphi(0)B_{21}.$$

Выберем полиномиальные функции

$$\varphi_1(z) = 2z - z^2; \quad \varphi_2(z) = z^2 - z; \quad \varphi_3(z) = (z-1)^2.$$

При этом получим компоненты матрицы A

$$B_{11} = 2A - A^2 = \begin{pmatrix} -2 & -3 & 3 \\ -4 & -3 & 4 \\ -6 & -6 & 7 \end{pmatrix} = A;$$

$$B_{12} = A^2 - A = \begin{pmatrix} 0 & 0 & 0 \\ 0 & 0 & 0 \\ 0 & 0 & 0 \end{pmatrix};$$

$$B_{21} = (A - E)^2 = \begin{pmatrix} 3 & 3 & -3 \\ 4 & 4 & -4 \\ 6 & 6 & -6 \end{pmatrix}.$$

В данном примере $B_{12} = 0$, т.е. матрица A имеет простую структуру. Любая аналитическая функция от матрицы A имеет вид

$$\varphi(A) = \varphi(1) \cdot \begin{pmatrix} -2 & -3 & 3 \\ -4 & -3 & 4 \\ -6 & -6 & 7 \end{pmatrix} + \varphi(0) \begin{pmatrix} 3 & 3 & -3 \\ 4 & 4 & -4 \\ 6 & 6 & -6 \end{pmatrix}.$$

Так при $\varphi(z) = z^n \quad (n = 1,2,3,...)$ получим

$$A^n = A, \quad (n = 1,2,...)$$

т.е. матрица A является проектором.

Замечание. Пусть матрица A имеет кратные собственные числа. Тогда аналитическая функция $\varphi(A)$ от матрицы A находится по формуле (61). При этом компоненты матрицы B_{k1} находятся по формуле

$$B_{k1} = \frac{1}{2\pi i} \int_{\Gamma_k} (Ez - A)^{-1} dz, \qquad (k = 1,\ldots,q)$$

и являются следовательно, проекторами матрицы A.

$$B_{k1} = P_k, \qquad (k = 1,\ldots,q).$$

контур Γ_k охватывает лишь точку $z = \lambda_k$ на комплексной плоскости z. Остальные компоненты матрицы A можно определить по формулам

$$B_{ks} = \frac{1}{(s-1)!} \cdot \frac{1}{2\pi i} \int_{\Gamma_k} (Ez - A)^{-1} (z - \lambda_k)^s dz,$$

$$(k = 1,\ldots,q; \; s = 1,\ldots,m_k).$$

(64)

Все компоненты B_{ks} матрицы A являются аналитическими функциями от матрицы A и поэтому все перестановочны

$$B_{ks} \cdot B_{ij} = B_{ij} \cdot B_{ks}, \qquad (k,i = 1,\ldots,q; \; s,j = 1,\ldots,m_k).$$

8. Подпространства

Для понимания следующих результатов необходимо использовать понятие подпространства.

Пусть L – некоторое множество векторов в m-мерном пространстве. Если любая линейная комбинация векторов из L тоже является вектором из L, то множество векторов L называется подпространством при выполнении следующих условий:

1. Если $X_1 \in L$, $X_2 \in L$, то $X_1 + X_2 \in L$.

2. Если $X \in L$, λ – произвольное число, то $\lambda X \in L$.

Нулевой вектор 0 принадлежит любому подпространству.

Можно дать другое определение подпространства L.

Пусть даны линейно независимые векторы $X_1, X_2, ..., X_q$.

Множество всех векторов вида

$$X = \alpha_1 X_1 + \alpha_2 X_2 + ... + \alpha_q X_q$$

при произвольных значениях коэффициентов $\alpha_1, \alpha_2, ..., \alpha_q$ образует подпространство L размерности q. Если в подпространстве L

существуют q линейно независимых векторов, а любые $q+1$ векторы L являются линейно зависимы, то целое число q называется размерностью подпространства и обозначается через $\dim L$.

Подпространство L, определяемое линейной комбинацией векторов $X_1, X_2, ..., X_q$, называется также подпространством, натянутым на векторы $X_1, X_2, ..., X_q$ или линейной оболочкой векторов $X_1, X_2, ..., X_q$.

Возьмем произвольный вектор X_0. Чтобы узнать, принадлежит ли вектор X_0 подпространству L, следует искать решение системы уравнений

$$\alpha_1 X_1 + \alpha_2 X_2 + ... + \alpha_q X_q = X_0 \qquad (65)$$

относительно неизвестных $\alpha_1, \alpha_2, ..., \alpha_q$. Если система уравнений (65) имеет решение, то $X_0 \in L$. Если система уравнений (65) не имеет решения, то $X_0 \overline{\in} L$. Линейно независимые векторы $X_1, X_2, ..., X_q$ называются базисом подпространства L, натянутого на векторы $X_1, X_2, ..., X_q$. В формуле (65) коэффициенты $\alpha_1, \alpha_2, ..., \alpha_q$ называются координатами вектора X_0.

Легко доказывается, что сумма $L_1 \cup L_2$ двух подпространств L_1, L_2 и пересечение $L_1 \cap L_2$ этих подпространств тоже являются подпространствами. При этом будет справедливо равенство [2]

$$\dim L_1 + \dim L_2 = \dim L_1 \cup L_2 + \dim L_1 \cap L_2 . \qquad (66)$$

Если L_1, L_2 – два подпространства в m-мерном векторном подпространстве L и не существует общего вектора, за исключением нулевого вектора, принадлежащего каждому подпространству L_1, L_2, то подпространство $L = L_1 \cup L_2$ называется прямой суммой подпространств L_1, L_2 и обозначается

$$L = L_1 + L_2 .$$

При этом любой вектор $X \in L$ единственным образом представляется в виде суммы двух векторов X_1, X_2:

$$X = X_1 + X_2, \qquad X_1 \in L_1, \quad X_2 \in L_2 . \qquad (67)$$

Базис подпространства L может быть получен объединением базисов подпространств L_1, L_2.

Для задания конкретных подпространств можно использовать понятие аннулируемого подпространства и области значений матрицы.

Пусть матрица Т размера $m \times m$ имеет ранг, равный q $(q \leq m)$. Множество векторов X, удовлетворяющих уравнению

$$TX = 0 \qquad (68)$$

образует подпространство, которое называется ядром матрицы Т, или аннулируемым подпространством и обозначается KerТ. Очевидно, будет выполнено равенство

$$\dim Ker\text{Т} = m - rang\text{Т} = m - q. \qquad (69)$$

Из матрицы Т можно выбрать q линейно независимых строк, объединить их в матрицу Т_1 размера $q \times m$ и записать уравнение (68) в виде

$$\text{Т}_1 X = 0.$$

Множество векторов X, представленное в виде

$$X = TY, \qquad rang\text{Т} = q, \qquad (70)$$

где Y – некоторый вектор размерности m, называется образом матрицы Т, или областью значений и обозначается через $\text{I}_m\,\text{Т}$. Образ матрицы Т является подпространством, базис которого образуется любыми q линейно независимыми вектор-столбцами матрицы Т. Справедливо очевидное равенство

$$\dim \text{I}_m\,\text{Т} = rang\text{Т}. \qquad (71)$$

Поэтому для любой квадратной матрицы T порядка m верна формула

$$\dim I_m T + \dim Ker T = m. \qquad (72)$$

В случае, когда $rang T = m$, получим

$$\dim I_m T = m, \quad \dim Ker T = 0.$$

При $0 < q < m$ можно выбрать q линейно независимых столбцов матрицы T, образовать из них матрицу T_2 размера $m \times q$ и задать вектор X (70) равенством

$$X = T_2 Z, \quad \dim Z = q. \qquad (73)$$

При $0 < q < m$ можно задать конкретное подпространство системой уравнений вида (68) или (70). Всегда можно перейти от уравнений вида (68) к уравнениям вида (70) и обратно.

Укажем простой способ перехода.

Пусть подпространство L задано системой уравнений вида (68). Выберем в матрице T q линейно независимых строк, образуем из них первые q строк вспомогательной матрицы T_3 размера $m \times m$, а остальные $m - q$ строк выбираем произвольным образом так, чтобы $\det T_3 \neq 0$.

Найдем обратную матрицу T_3^{-1}. Поскольку $T_3 T_3^{-1} = E$, то в качестве матрицы T_2 можно взять матрицу, содержащую последние $m-q$ столбцы матрицы T_3^{-1}. При этом подпространство L_2 можно представить в виде системы уравнений (73).

Действительно в силу построения $T_1 T_2 = 0$.

Пример. Пусть задана матрица

$$T = \begin{pmatrix} 1 & -1 & 2 \\ 2 & -2 & 4 \\ -1 & 1 & -2 \end{pmatrix}, \quad rang T = 1.$$

Образуем матрицу T_3 и найдем T_3^{-1}

$$T_3 = \begin{pmatrix} 1 & -1 & 2 \\ 0 & 1 & 0 \\ 0 & 0 & 1 \end{pmatrix}, \quad T_3^{-1} = \begin{pmatrix} 1 & 1 & -2 \\ 0 & 1 & 0 \\ 0 & 0 & 1 \end{pmatrix}.$$

Из двух последних столбцов матрицы T_3^{-1} образуем матрицу T_2. Подпространство $TX = 0$ можно записать в виде

$$X = \begin{pmatrix} 1 & -2 \\ 1 & 0 \\ 0 & 1 \end{pmatrix} Z, \quad \dim Z = 2.$$

Пусть наоборот подпространство L задано системой уравнений вида (70). Выделим q линейно независимых столбцов матрицы T и образуем из них первые q столбцов матрицы T_4 размера $m \times m$.

Последние $m - q$ столбцов выберем произвольно так, чтобы $\det T_4 \neq 0$. Поскольку $T_4^{-1}T_4 = E$, то в качестве матрицы T_1 можно взять нижние $m - q$ строк матрицы T_4^{-1} и записать подпространство в виде системы уравнений (68).

Пример. Пусть подпространство L задано системой уравнений $X = TY$, где матрица T задана формулой

$$T = \begin{pmatrix} 1 & -1 & 2 \\ 2 & -2 & 4 \\ -1 & 1 & -2 \end{pmatrix}, \quad rang\,T = 1.$$

Образуем матрицу T_4 и найдем матрицу T_4^{-1}

$$T_4 = \begin{pmatrix} 1 & 0 & 0 \\ 2 & 1 & 0 \\ -1 & 0 & 1 \end{pmatrix}, \quad T_4^{-1} = \begin{pmatrix} 1 & 0 & 0 \\ -2 & 1 & 0 \\ 1 & 0 & 1 \end{pmatrix}.$$

В качестве матрицы T_1 можно взять нижние две строки матрицы T_4^{-1} и определить подпространство L системой уравнений

$$\begin{pmatrix} -2 & 1 & 0 \\ 1 & 0 & 1 \end{pmatrix} X = 0.$$

9. Инвариантные подпространства матрицы, подобной блочной матрице

Пусть матрица A подобна блочной диагональной матрице Λ, т.е. существует матрица T такая, что

$$T^{-1}AT = \Lambda, \quad \Lambda = \begin{pmatrix} \Lambda_1 & 0 & \ldots & 0 \\ 0 & \Lambda_2 & \ldots & 0 \\ \ldots & \ldots & \ldots & \ldots \\ 0 & 0 & \ldots & \Lambda_q \end{pmatrix}. \quad (74)$$

Матрицы Λ_k – квадратные матрицы порядка m_k $\left(m_1 + m_2 + \ldots + m_q\right) = m$. Разобьем матрицы T, T^{-1} на блоки порядков соответственно $m \times m_k$, $m_k \times m$

$$T = \begin{pmatrix} T_1 & T_2 & \ldots & T_q \end{pmatrix}, \quad T^{-1} = \begin{pmatrix} U_1 \\ U_2 \\ \ldots \\ U_q \end{pmatrix}.$$

Пусть функция $\varphi(z)$ аналитична в некоторой области D, содержащей спектр матрицы A. Предположим, что спектры матриц Λ_k отделяются и могут быть заключены в простые непересекающиеся контуры Γ_k. Пусть контур Γ охватывает спектр матрицы A. Найдем функцию $\varphi(A)$ по формуле

$$\varphi(A) = \frac{1}{2\pi i} \int_{\tilde{A}} \varphi(z) \, (Ez - A)^{-1} \, dz =$$

$$= \frac{1}{2\pi i} \int_{\tilde{A}} \varphi(z) \, \left(Ez - T\Lambda T^{-1}\right)^{-1} \, dz = \qquad (75)$$

$$= T \frac{1}{2\pi i} \int_{\tilde{A}} \varphi(z) \, (Ez - \Lambda)^{-1} \, dz \cdot T^{-1}.$$

Поскольку для блочной матрицы справедливо равенство

$$(Ez - \Lambda)^{-1} = \begin{pmatrix} (Ez - \Lambda_1)^{-1} & 0 & \ldots & 0 \\ 0 & (Ez - \Lambda_2)^{-1} & \ldots & 0 \\ \ldots & \ldots & \ldots & \ldots \\ 0 & 0 & \ldots & (Ez - \Lambda_q)^{-1} \end{pmatrix},$$

то получим для функции от матрицы A

$$\varphi(A) = T \begin{pmatrix} f(\Lambda_1) & 0 & \ldots & 0 \\ 0 & f(\Lambda_2) & \ldots & 0 \\ \ldots & \ldots & \ldots & \ldots \\ 0 & 0 & \ldots & f(\Lambda_q) \end{pmatrix} T^{-1}$$

известную формулу для функции от блочной матрицы

$$\varphi(A) = T_1 \varphi(\Lambda_1) U_1 + T_2 \varphi(\Lambda_2) U_2 + \ldots + T_q \varphi(\Lambda_q) U_q. \qquad (76)$$

Пример. Пусть матрица A – диагональная

$$A = \begin{pmatrix} \alpha_1 & 0 & ... & 0 \\ 0 & \alpha_2 & ... & 0 \\ ... & ... & ... & ... \\ 0 & 0 & ... & \alpha_m \end{pmatrix}.$$

Тогда имеем равенство

$$\varphi(A) = \begin{pmatrix} \varphi(\alpha_1) & 0 & ... & 0 \\ 0 & \varphi(\alpha_2) & ... & 0 \\ ... & ... & ... & ... \\ 0 & 0 & ... & \varphi(\alpha_m) \end{pmatrix}.$$

В частном случае получим равенство

$$e^{At} = \begin{pmatrix} e^{\alpha_1 t} & 0 & ... & 0 \\ 0 & e^{\alpha_2 t} & ... & 0 \\ ... & ... & ... & ... \\ 0 & 0 & ... & e^{\alpha_m t} \end{pmatrix}. \tag{77}$$

Если матрица A подобна диагональной матрице

$$T^{-1}AT = \begin{pmatrix} \alpha_1 & 0 & ... & 0 \\ 0 & \alpha_2 & ... & 0 \\ ... & ... & ... & ... \\ 0 & 0 & ... & \alpha_m \end{pmatrix},$$

то для показательной матрицы получим выражение

$$e^{At} = T \begin{pmatrix} e^{\alpha_1 t} & 0 & \dots & 0 \\ 0 & e^{\alpha_2 t} & \dots & 0 \\ \dots & \dots & \dots & \dots \\ 0 & 0 & \dots & e^{\alpha_m t} \end{pmatrix} T^{-1} . \qquad (78)$$

Любой элемент матрицы e^{At} является линейной комбинацией показательных функций

$$e^{At} = \sum_{k=1}^{m} T_k e^{\alpha_k t} U_k , \qquad (79)$$

где T_k, U_k – постоянные матрицы.

Для блочной матрицы найдем выражения для проекторов матрицы A по формулам

$$P_k = \frac{1}{2\pi i} \int_{\Gamma_k} \left(Ez - A \right)^{-1} dz = \frac{1}{2\pi i} \int_{\Gamma_k} \sum_{s=1}^{m} T_s \left(Ez - \Lambda_s \right)^{-1} U_s dz =$$

$$= T_k \frac{1}{2\pi i} \int_{\Gamma_k} \left(Ez - \Lambda_k \right)^{-1} dz\, U_k = T_k E_{m_k} U_k = T_k U_k \ \left(k = 1, 2, \dots, q \right).$$

$$(80)$$

Здесь E_{m_k} – единичная матрица порядка m_k.

Проекторы P_k $\left(k = 1, 2, \dots, q \right)$ определяют линейные подпространства $L_k = I_m\, P_k$. при этом ядром матрицы P_k будет подпространство столбцовых векторов, ортогональных к столбцам матрицы U_k^*, т.е. образ матрицы $E - P_k$.

Аналогично линейные подпространства L_k^* $(k = 1,2,...,q)$ пространства строчных векторов L^* образованы линейными комбинациями строк матрицы U_k $(k = 1,2,...,q)$.

Ранг проектора P_k равен порядку m_k матрицы Λ_k. Из формулы

$$P_k = \mathrm{T}_k U_k \qquad (k = 1,2,...,q) \qquad (81)$$

следует, что столбцы проектора P_k являются линейными комбинациями столбцов матрицы T_k, а строки проектора P_k являются линейными комбинациями строк матрицы U_k.

Если известен проектор P_k, то базис подпространства можно найти, выделяя линейно независимые столбцы матрицы P_k. Аналогично базис подпространства L_k^* можно найти, выделяя линейно независимые строки матрицы P_k. Из условия

$$U_k \mathrm{T}_k = \mathrm{E}_{m_k} \qquad (k = 1,2,...,q)$$

находим разложение проектора P_k на множители вида (81).

Покажем теперь, что подпространства L_k $(k = 1,2,...,q)$, определяемые проекторами P_k

$(k = 1,2,..,q)$ являются инвариантными подпространствами матрицы A по формуле (75) при $\varphi(z) \equiv z$:

$$A = \sum_{k=1}^{q} \frac{1}{2\pi i} \int\limits_{\tilde{A}_k} z(Ez - A)^{-1} dz = \sum_{k=1}^{q} \frac{1}{2\pi i} \int\limits_{\tilde{A}_k} T_k z (Ez - \Lambda_k)^{-1} U_k dz =$$
$$= \sum_{k=1}^{q} \frac{1}{2\pi i} \int\limits_{\tilde{A}_k} T_k \Lambda_k (Ez - \Lambda_k)^{-1} U_k dz = \sum_{k=1}^{q} T_k \Lambda_k U_k$$
.(82)

Из очевидных формул

$$U_k T_k = E_{m_k}, \qquad U_k T_s = 0 \qquad (k \neq s; \; k,s = 1,2,...,q), \quad (83)$$

следующих из равенства $T^{-1}T = E$ вытекает, что при $X \in L_k$ будет выполнено соотношение

$$AX = \sum_{k=1}^{q} T_k \Lambda_k U_k X = T_k (\Lambda_k U_k X), \qquad AX \in L_k.$$

Следовательно, подпространство L_k, являющееся образом проектора P_k, будет инвариантным подпространством матрицы A.

Из формул (82), (83) вытекают равенства

$$AT_k = T_k \Lambda_k, \qquad U_k A = \Lambda_k U_k \qquad (k = 1,2,...,q), \qquad (84)$$

которые обобщают понятия собственного вектора и собственного числа матрицы. Матрицы T_k, U_k называем собственными матрицами, а матрицы Λ_k называем матричными множителями.

10. Инвариантные подпространства матрицы с кратными собственными значениями

Рассмотрим случай, когда в матрицах Λ_k все собственные числа одинаковы и равны λ_k.

Пусть матрицы Λ_k являются клетками Жордана:

$$\Lambda_k = \begin{pmatrix} \lambda_k & 1 & 0 & \ldots & 0 \\ 0 & \lambda_k & 1 & \ldots & 0 \\ 0 & 0 & \lambda_k & \ldots & 0 \\ \ldots & \ldots & \ldots & \ldots & \ldots \\ 0 & 0 & 0 & \ldots & \lambda_k \end{pmatrix} \qquad (85)$$

размера $m_k \times m_k$. Различные матрицы Λ_k могут иметь одинаковые собственные числа. Обозначим через $X_{k1}, \ldots, X_{km_k}$ столбцы матрицы T_k. Из первого матричного равенства (84) вытекают равенства

$$AX_{k1} = \lambda_k X_{k1}, \quad AX_{k2} = \lambda_k X_{k2} + X_{k1},$$

$$AX_{km_k} = \lambda_k X_{km_k} + X_{km_k - 1}. \qquad (86)$$

Вектор X_{k1} называется собственным вектором, а векторы $X_{k2}, \ldots, X_{km_k}$ называются присоединенными к X_{k1} векторами.

Векторы $X_{k1}, X_{k2}, \ldots, X_{km_k}$ порождающие инвариантное подпространство L_k матрицы A, соответствуют клетке Жордана Λ_k. Из равенства (86) получим формулы

$$\left(A - \lambda_k E\right)X_{k1} = 0, \ \left(A - \lambda_k E\right)^2 X_{k2} = 0, \ ..., \ \left(A - \lambda_k E\right)^{m_k} X_{km_k} = 0.$$
$$(87)$$

Следовательно, векторы $X_{k1}, X_{k2}, ..., X_{km_k}$ входят в ядро матрицы $\left(A - \lambda_k\right)^{m_k}$. Поэтому справедливо равенство

$$\left(A - \lambda_k E\right)^{m_k} T_k = 0. \qquad (88)$$

Если несколько клеток Жордана имеют одинаковые собственные числа, то все векторы базисов инвариантных подпространств удовлетворяют векторному уравнению

$$\left(A - \lambda_k E\right)^{v_k} X = 0, \qquad (89)$$

где v_k – наибольший из порядков клеток Жордана, имеющих собственное число λ_k.

Аналогично, строчные векторы, которые образуют базис инвариантного подпространства в сопряженном пространстве L^* удовлетворяют уравнению

$$U_k \left(A - \lambda_k E\right)^{m_k} = 0$$

Все вектора Y базисов инвариантных подпространств, порождаемых клетками Жордана с одинаковыми собственными λ_k, удовлетворяет векторному уравнению

$$Y\left(A - \lambda_k E\right)^{\nu_k} = 0. \qquad (90)$$

Рассмотрим подробнее строение аналитической функции от матрицы в случае кратных корней характеристического уравнения. Предполагаем, что в равенстве (74) все матрицы Λ_{π} являются клетками Жордана (85). Положим

$$\Lambda_k = E_{m_k} \lambda_k + F_{m_k}, \qquad (91)$$

где матрица F_{m_k} имеет порядок m_k. Все элементы матрицы F_{m_k} равны нулю, за исключением единичных элементов, занимающих одну диагональ выше главной:

$$F_{m_k} = \begin{pmatrix} 0 & 1 & 0 & ... & 0 \\ 0 & 0 & 1 & ... & 0 \\ ... & ... & ... & ... & ... \\ 0 & 0 & 0 & ... & 1 \\ 0 & 0 & 0 & ... & 0 \end{pmatrix}. \qquad (92)$$

По формуле (61) получим аналитическое выражение для функции от матрицы

$$\varphi(A) = \sum_{k=1}^{q} T \frac{1}{2\pi i} \int_{\tilde{A}_k} \varphi(z)(Ez - \Lambda)^{-1} dz T^{-1} =$$

$$= \sum_{k=1}^{q} T_k \frac{1}{2\pi i} \int_{\tilde{A}_k} \varphi(z)\left(E_{m_k} z - \lambda_k E_{m_k} - F_{m_k}\right)^{-1} dz \ U_k .$$

(93)

Поскольку справедливо разложение

$$\left(E_{m_k}(z - \lambda_k) - F_{m_k}\right)^{-1} = \frac{1}{z - \lambda_k} E_{m_k} + \frac{1}{(z - \lambda_k)^2} F_{m_k} +$$

$$+ \frac{1}{(z - \lambda_k)^3} F_{m_k}^2 + ... + \frac{1}{(z - \lambda_k)^{m_k - 1}} F_{m_k}^{m_k - 1},$$

(94)

то из формулы (93) получим аналитическое представление функции от матрицы A

$$\varphi(A) = \sum_{k=1}^{q} T_k (\varphi(\lambda_k) E_{m_k} + \frac{\varphi'(\lambda_k)}{1!} F_{m_k} + \frac{\varphi''(\lambda_k)}{2!} F_{m_k}^2 + .$$

$$+ ... + \frac{\varphi^{(m_k-1)}(\lambda_k)}{(m_k - 1)!} F_{m_k}^{m_k - 1}) U_k$$

(95)

Сопоставление с формулой (61) позволяет получить явное выражение для компонентов B_{ks} матрицы A

$$B_{k1} = T_k U_k , \quad B_{ks} = T_k F_{m_k}^{s-1} U_k \quad (s = 2,...,m_k).$$

(96)

Из этих формул видно, что

$$I_m B_{ks} \subset I_m B_{k1} \quad (s = 2,...,m_k).$$

(97)

Если положим по определению, что $F_{m_k}^0 = E_{m_k}$, то формулы (96) можно объединить в одну формулу

$$B_{ks} = T_k F_{m_k}^{s-1} U_k \qquad \left(s = 1,2,\ldots,m_k\right). \qquad (98)$$

Поскольку $U_k T_k = E_{m_k}$, $U_k T_s = 0$ $\left(k \neq s\right)$, то получим равенства для компонентов матрицы A

$$B_{ks} B_{nj} = T_k F_{m_k}^{s-1} U_k T_n F_{m_k}^{j-1} U_n = \delta_{kn} T_k F_{m_k}^{s+j-2} U_k = \delta_{kn} B_{k,s+j-1}.$$

Следует учесть, что $F_{m_k}^{m_k} = 0$, т.е. $B_{ks} = 0$ $\left(s > m_k\right)$.

Многие свойства функций от матрицы изложены в работе [1].

11. Алгебраический способ построения проекторов

Приведем способ построения проекторов матрицы в случае, когда известен минимальный многочлен

$$\psi(z) = (z - \lambda_1)^{v_1} (z - \lambda_2)^{v_2} \dots (z - \lambda_q)^{v_q}.$$

Минимальный многочлен - это многочлен наименьшей степени такой, что $\psi(A) = 0$.

Введем в рассмотрение вспомогательные многочлены

$$\psi_k(z) = \frac{\psi(z)}{(z - \lambda_k)^{v_k}} \qquad (k = 1,2,\dots,q). \qquad (99)$$

Поскольку многочлены $\psi_k(z)$ $(k = 1,2,\dots,q)$ являются взаимно простыми, то найдутся многочлены $f_k(z)$ удовлетворяющие тождеству

$$f_1(z)\,\psi_1(z) + f_2(z)\,\psi_2(z) + \dots + f_q(z)\,\psi_q(z) \equiv 1.$$

Перейдем в этом тождестве от скалярного аргумента z к матричному аргументу A. Получим матричное равенство

$$f_1(A)\,\psi_1(A) + f_2(A)\,\psi_2(A) + \dots + f_q(A)\,\psi_q(A) = E.$$

Покажем, что слагаемые в этом равенстве

$$P_k \equiv f_k(A)\,\psi_k(A) \qquad (100)$$

являются проекторами, которые определяются собственными числами λ_k матрицы A. Поскольку многочлен

$$\left(z - \lambda_k\right)^{v_k} f_k\left(z\right) \psi_k\left(z\right) = \psi\left(z\right) f_k\left(z\right)$$

делится нацело на минимальный многочлен $\psi\left(z\right)$, то будут справедливы равенства

$$\left(\mathrm{A} - \lambda_k \mathrm{E}\right)^{v_k} f_k\left(\mathrm{A}\right) \psi_k\left(\mathrm{A}\right) = \left(\Lambda - \lambda_k \mathrm{E}\right)^{v_k} P_k \quad \left(k = 1, 2, \ldots, q\right).$$

(101)

Следовательно, столбцы матрицы P_k входят в инвариантное подпространство L_k матрицы A, соответствующее собственному числу λ_k.

Поскольку многочлен $f_k\left(z\right) \psi_k\left(z\right) f_s\left(z\right) \psi_s\left(z\right)$ $\left(k \neq s\right)$ делится нацело на минимальный многочлен, то

$$f_k\left(\mathrm{A}\right) \psi_k\left(\mathrm{A}\right) f_s\left(z\right) \psi_s\left(\mathrm{A}\right) = P_k P_s = 0 \quad \left(k \neq s\right).$$

(102)

Наконец, умножая равенство

$$P_1 + P_2 + \ldots + P_q = \mathrm{E},$$

получаемое из (98), на P_k, получим в силу формул (102) равенство $P_k P_k = P_k$. Отсюда следует, что матрицы P_k $\left(k = 1, 2, \ldots, q\right)$ (100) образуют полную группу проекторов.

Замечание. Если аннулирующий многочлен $\psi(z)$ матрицы A порядка m разбит на два взаимно простых множителя

$$\psi(z) = \psi_1(z) \cdot \psi_2(z), \qquad (103)$$

то все m-мерное пространство L можно соответственно разложить в прямую сумму двух подпространств L_1, L_2, которые являются инвариантными подпространствами матрицы A. При этом будет справедливо разложение

$$X = X_1 + X_2, \quad X_k \in L_k \quad (k = 1,2);$$
$$\psi_1(A)X_2 = 0, \quad \psi_2(A)X_1 = 0.$$
$$(104)$$

Для построения проекторов P_1, P_2 найдем многочлены $f_1(z)$, $f_2(z)$ такие, что

$$f_1(z)\ \psi_1(z) + f_2(z)\ \psi_2(z) = 1. \qquad (105)$$

Аналогично можно доказать, что проекторы, определяющие разложение (104), имеют вид

$$P_1 = f_1(A)\ \psi_1 A, \quad P_2 = f_2(A)\ \psi_2(A). \qquad (106)$$

Пример. Найдем проекторы матрицы A

$$A = \begin{pmatrix} 1 & 2 \\ 4 & 3 \end{pmatrix}.$$

Составим характеристический многочлен, которой разлагается на два множителя

$$\begin{vmatrix} z-1 & -2 \\ -4 & z-3 \end{vmatrix} = (z+1)\ (z-5).$$

Из тождества

$$\frac{1}{6}(z+1)-\frac{1}{6}(z-5)\equiv 1$$

находим проекторы матрицы A:

$$P_1 = \frac{1}{6}(A+E) = \frac{1}{3}\begin{pmatrix} 1 & 1 \\ 2 & 2 \end{pmatrix},$$

$$P_2 = -\frac{1}{6}(A-5E) = \frac{1}{3}\begin{pmatrix} 2 & 1 \\ -2 & 1 \end{pmatrix}.$$

Приложения теории линейных проекторов для отыскания собственных чисел матрицы изложены в работе [1], где введены нелинейные проекторы $P_j(t,X)$ $(j=1,2)$, удовлетворяющие свойствам:

$$P_j(t,P_j(t,X))\equiv P_j(t,X),\quad P_j(t,P_s(t,X))\equiv 0$$

$$(j\neq s,\ j,s=1,2),$$

$$P_1(t,X)+P_2(t,X)\equiv X,\quad P_j(t,0)\equiv 0. \qquad (107)$$

Нелинейные проекторы позволяют расщеплять решения нелинейной системы дифференциальных уравнений.

12. Приложения теории матриц к дифференциальным и разностным уравнениям

Система линейных дифференциальных уравнений

$$\frac{dX}{dt} = AX \qquad (108)$$

имеет фундаментальную матрицу решений

$$N(t) = e^{At}. \qquad (109)$$

Решение задачи Коши $t = t_0$, $X = X_0$ имеет вид

$$X(t) = e^{A(t-t_0)}X(t_0).$$

Пример. Найдем фундаментальную матрицу решений системы дифференциальных уравнений (108), где

$$A = \begin{pmatrix} 1 & 2 \\ 4 & 3 \end{pmatrix}, \quad \lambda_1 = -1, \quad \lambda_2 = 5. \qquad (110)$$

Находим проекторы и общий вид функции от матрицы A

$$P_1 = \frac{1}{3}\begin{pmatrix} 2 & -1 \\ -2 & 1 \end{pmatrix}, \quad P_2 = \frac{1}{3}\begin{pmatrix} 1 & 1 \\ 2 & 2 \end{pmatrix},$$

$$\varphi(A) = P_1\varphi(\lambda_1) + P_2\varphi(\lambda_2).$$

Для функции $\varphi(z) = e^{zt}$ получим

$$e^{At} = \frac{1}{3}\begin{pmatrix} 2 & -1 \\ -2 & 1 \end{pmatrix}e^{-t} + \frac{1}{3}\begin{pmatrix} 1 & 1 \\ 2 & 2 \end{pmatrix}e^{5t} = \begin{pmatrix} \dfrac{e^{5t} + 2e^{-t}}{3} & \dfrac{e^{5t} - e^{-t}}{3} \\ \dfrac{2e^{5t} - 2e^{-t}}{3} & \dfrac{2e^{5t} + e^{-t}}{3} \end{pmatrix},$$

фундаментальную матрицу решений.

Аналогично может быть решена система линейных разностных уравнений

$$X_{n+1} = AX_n \qquad (n = 0,1,2,...). \tag{111}$$

Эта система уравнений имеет общее решение

$$X_n = A^n X_0 \qquad (n = 0,1,2,...).$$

Пример. Найдем решение системы разностных уравнений (111), где матрица A дана в формуле (110). Для функции $\varphi(z) = z^n$ получим

$$A^n = P_1(-1)^n + P_2(5^n) = \frac{1}{3}\begin{pmatrix} 2 & -1 \\ -2 & 1 \end{pmatrix}(-1)^n + \frac{1}{3}\begin{pmatrix} 1 & 1 \\ 2 & 2 \end{pmatrix}5^n.$$

Общее решение системы имеет вид

$$X_n = \frac{1}{3}\begin{pmatrix} 5^n + 2\cdot(-1)^n & 5^n + (-1)^n \\ 2\cdot 5^n - 2\cdot(-1)^n & 2\cdot 5^n + (-1)^n \end{pmatrix}X_0$$

$$(n = 0,1,2,...).$$

Аналогично решаются системы уравнений порядка выше первого. Ищем решение системы дифференциальных уравнений

$$\frac{d^m Y(t)}{dt^m} + \sum_{s=0}^{m-1} A_s \frac{d^s Y(t)}{dt^s} = 0$$

в виде

$$Y(t) = e^{\lambda t} C .$$

При этом приходим к системе уравнений

$$\left(E\lambda^m + \sum_{s=0}^{m-1} A_s \lambda^s \right) C = 0 .$$

Показатель λ удовлетворяет уравнению

$$\det\left(E\lambda^m + \sum_{s=0}^{m-1} A_s \lambda^s \right) = 0 . \qquad (112)$$

Для каждого корня $\lambda = \lambda_j$, уравнения (112) находим соответствующий вектор C_j. Если все корни уравнения (112) различны, то находим общее решение системы дифференциальных уравнений

$$Y(t) = \sum C_j e^{\lambda_j t} .$$

13. Основы метода усреднения

Метод усреднения создан классиками естествознания для решения задач небесной механики.

Этот метод усреднения приобрел особую популярность среди украинских математиков и тесно связан с работами Н. Н. Боголюбова, Ю. А. Митропольского, И. З. Штокало и других.

Принципиально новые результаты по методу усреднения получены в наших работах [3,5].

 В этих работах объясняется глубинная суть метода усреднения, а именно – связь метода усреднения с нормальными формами системы дифференциальных уравнений.

Определение. Систему дифференциальных уравнений

$$\frac{dX}{dt} = \mathrm{A}X + F(X) \qquad (113)$$

будем называть приведенной к нормальной форме относительно матрицы A, если выполнено тождество

$$e^{\mathrm{A}t} F\left(e^{-\mathrm{A}t} X\right) \equiv F(X), \quad -\infty < t < \infty. \qquad (114)$$

Замена переменных

$$X = e^{\mathrm{A}t} Y$$

преобразует систему уравнений (113) в систему уравнений

$$\frac{dY}{dt} = F(Y).$$

Приведенное нами определение нормальной формы совпадает с существующим определением нормальной формы [4].

Пусть A – диагональная матрица с собственными числами $\alpha_1, \alpha_2, ..., \alpha_m$. При этом имеем равенство

$$e^{At} = \begin{pmatrix} e^{\alpha_1 t} & 0 & ... & 0 \\ 0 & e^{\alpha_2 t} & ... & 0 \\ ... & ... & ... & ... \\ 0 & 0 & ... & e^{\alpha_m t} \end{pmatrix}.$$

Если обозначим

$$F(X) = \begin{pmatrix} f_1(x) \\ f_2(x) \\ ... \\ f_m(x) \end{pmatrix}, \tag{115}$$

то из условия (114) получим, что

$$f_j(X) = x_j \varphi_j(X) \quad (j = 1,2,...,m), \tag{116}$$

где $\varphi_j(X)$ – интегралы системы уравнений

$$\frac{dX}{dt} = \mathrm{A}X\,,$$

т.е. выполнены тождества

$$\varphi_j\left(x_1 e^{\alpha_1 t}, x_2 e^{\alpha_2 t}, ..., x_m e^{\alpha_m t}\right) \equiv \varphi_j\left(x_1, x_2, ..., x_m\right).$$

Пример. Если система дифференциальных уравнений

$$\frac{dx}{dt} = \alpha x + f_1(x, y), \quad \frac{dy}{dt} = -\alpha y + f_2(x, y), \quad \alpha \neq 0$$

приведена к нормальной форме, то она примет вид

$$\frac{dx}{dt} = \alpha x + x\varphi_1(xy), \quad \frac{dy}{dt} = -\alpha y + y\varphi_2(xy).$$

Приведение к нормальной форме осуществляется с помощью обобщения операции усреднения. А именно, полагаем

$$\left\langle t^n e^{\alpha t}\right\rangle = 0 \quad \left(n = 0, 1, 2, ...; \ \alpha \neq 0\right) \qquad (117)$$

$$\left\langle c\right\rangle = c\,, \quad c = const\,,$$

$\left\langle t^n\right\rangle$ – не существует при $n = 1, 2, 3, ...$.

Если $y(t)$ – функция, имеющая изображение по Лапласу

$$f(p) = \int_0^\infty e^{-pt} y(t) dt\,,$$

то полагаем [5]

$$\langle y(t) \rangle = \lim_{p \to 0} p f(p). \qquad (118)$$

Рассматривается система дифференциальных уравнений

$$\frac{dZ}{dt} = AZ + \mu F(Z, \mu), \qquad (119)$$

где вектор-функция $F(Z, \mu)$ разлагается в формальные степенные ряды по степеням z_j, μ.

После линейной замены

$$Z = e^{At} X \qquad (120)$$

система уравнений (119) преобразуется в нестационарную систему дифференциальных уравнений

$$\frac{dX}{dt} = \mu e^{-At} F(e^{At} X, \mu), \qquad (121)$$

которая не имеет нормальной формы относительно матрицы A.

Пусть построена замена переменных

$$X = Y + \mu \Psi(t, Y, \mu); \quad \langle \Psi(t, Y, \mu) \rangle = 0, \qquad (122)$$

преобразующая систему уравнений (121) в стационарную систему дифференциальных уравнений

$$\frac{dY}{dt} = \mu \Pi(Y, \mu), \qquad (123)$$

правая часть которой не зависит явно от времени t. Замена переменных

$$U = e^{At}Y$$

преобразует систему уравнений (123) в систему уравнений

$$\frac{dU}{dt} = AU + \mu e^{At}\Pi\left(e^{-At}U,\mu\right). \qquad (124)$$

Вектор-функция $\Pi(U,\mu)$ имеет нормальную форму относительно матрицы A, т.е. справедливо тождество

$$e^{At}\Pi\left(e^{-At}U,\mu\right) \equiv \Pi(U,\mu).$$

Замена переменных

$$Z = U + \mu e^{At}\Psi\left(t,e^{-At}U,\mu\right)$$

не зависит фактически от времени в силу тождества

$$e^{At}\Psi\left(t,e^{-At}U,\mu\right) \equiv \Psi(0,U,\mu).$$

Отсюда следует, что замена переменных

$$Z = U + \mu\Psi(0,U,\mu)$$

является нормализующей, т.е. преобразует исходную систему уравнений (119) в систему уравнений

$$\frac{dU}{dt} = AU + \mu\Pi(U,\mu), \qquad (125)$$

имеющую нормальную форму.

Справедливость приведенных утверждений вытекает из следующей основной теоремы [5].

Теорема. Если для системы дифференциальных уравнений

$$\frac{dX}{dt} = \mu e^{-At} F\left(e^{At} X, \mu\right), \qquad (126)$$

где $F(X, \mu)$ разлагается в формальные степенные ряды по степеням $x_1, \ldots, x_m$, μ построена замена

$$X = Y + \mu \Psi(t, Y, \mu); \qquad \langle \Psi(t, Y, \mu) \rangle \equiv 0, \qquad (127)$$

приводящая систему уравнений (126) к виду

$$\frac{dY}{dt} = \mu \Pi(y, \mu), \qquad (128)$$

то для замены (127) и системы уравнений (128) выполняются следующие свойства:

$$e^{A\tau} \Pi\left(e^{-A\tau} U, \mu\right) \equiv \Pi(U, \mu);$$
$$e^{A\tau} \Psi\left(t, e^{-A\tau} U, \mu\right) \equiv \Psi(t - \tau, U, \mu). \qquad (129)$$

При доказательстве теоремы используется операция интегрирования с нулевым средним.

$$\{Y(t)\} = \int_0^t \left(Y(t) - \langle Y(t) \rangle\right) dt - \left\langle \int_0^t \left(Y(t) - \langle Y(t) \rangle\right) dt \right\rangle, \qquad (130)$$

т.е. перед интегрированием функция усредняется и после интегрирования усредняется.

Пример. Приведем примеры применения операции интегрирования с нулевым средним для некоторых функций:

$$\left\{e^{2t}+1\right\}=\frac{1}{2}e^{2t};\quad\left\{\sin^2 t\right\}=\left\{\frac{1-\cos 2t}{2}\right\}=\frac{-\sin 2t}{4};$$

$$\left\{e^{3t}+e^{4t}+e^{5t}+6\right\}=\frac{1}{3}e^{3t}+\frac{1}{4}e^{4t}+\frac{1}{5}e^{5t};$$

$$\left\{sh^2 t\right\}=\frac{e^{2t}}{8}-\frac{e^{-2t}}{8}.$$

Выполним замену (127) в системе уравнений (126) и придем к системе уравнений с частными производными первого порядка

$$\mu\Pi(Y,\mu)+\mu\frac{\partial\Psi(t,Y,\mu)}{\partial t}+\mu^2\frac{D\Psi(t,Y,\mu)}{DY}\Pi(Y,\mu)=$$

$$=\mu e^{-At}F\left(e^{At}Y+\mu e^{At}\Psi(t,Y,\mu),\mu\right).$$

$$(131)$$

Используем операцию усреднения

$$\Pi(Y,\mu)=\left\langle e^{-At}F\left(e^{At}Y+\mu e^{At}\Psi(t,Y,\mu)\right)\right\rangle \qquad (132)$$

и операцию интегрирования с усреднением

$$\Psi(t,Y,\mu)=\left\langle e^{-At}F\left(e^{At}Y+\mu e^{At}\Psi(t,Y,\mu),\mu\right)-\right.$$

$$\left.-\mu\frac{D\psi(t,Y,\mu)}{DY}\Pi(Y,\mu)\right\rangle \qquad (133)$$

Решение системы (132), (133) можно искать, используя разложения по степеням параметра μ или методом последовательных приближений. При использовании метода последовательных приближений при

$$\Pi_0(Y,\mu) \equiv 0, \qquad \Psi_0(t,Y,\mu) \equiv 0$$

получим

$$\Pi_{k+1}(Y,\mu) = \left\langle e^{-At}F\left(e^{At}Y + \mu e^{At}\Psi_k(t,Y,\mu),\mu\right)\right\rangle;$$

$$\Psi_{k+1}(t,Y,\mu) = \left\{ e^{-At}F\left(e^{At}Y + \mu e^{At}\Psi_k(t,Y,\mu),\mu\right) - \right.$$

$$\left. -\mu\frac{D\Psi_k(t,Y,\mu)}{DY}\Pi_k(Y,\mu)\right\} \quad (k=0,1,2,\ldots).$$

$$(134)$$

При этом получим асимптотические соотношения

$$\left\|\Pi_k(Y,\mu) - \Pi(Y,\mu)\right\| = O(\mu^k);$$

$$\left\|\Psi_k(t,Y,\mu) - \Psi(t,Y,\mu)\right\| = O(\mu^k) \quad (k=0,1,2,\ldots),$$

$$(135)$$

т.е. первые члены разложений вектор-функций $\Pi(Y,\mu)$, $\Psi(t,Y,\mu)$ по степеням μ можно получить с точностью до μ^k включительно из разложений по степеням μ приближений $\Pi_k(Y,\mu)$, $\Psi_k(t,Y,\mu)$.

Рассмотрим систему дифференциальных уравнений

$$\frac{dZ_1}{dt} = A_1 Z_1 + \mu F_1 \left(Z_1, Z_2, \mu \right);$$

$$\frac{dZ_2}{dt} = A_2 Z_2 + \mu F_2 \left(Z_1, Z_2, \mu \right), \quad (136)$$

$$F_k \left(0, 0, \mu \right) \equiv 0 \quad \left(k = 1, 2 \right).$$

Предполагаем, что вектор-функции $F_k \left(Z_1, Z_2, \mu \right)$ достаточное число раз дифференцируемы по всем аргументам.

Пусть собственные числа матрицы A_1 лежат на мнимой оси, матрица A_1 имеет простую структуру. Собственные числа матрицы A_2 имеют отрицательные вещественные части.

Используя предложенный асимптотический метод преобразуем систему уравнений (136) в систему вида

$$\frac{dU_1}{dt} = A_1 U_1 + \mu \Pi_1 \left(U_1, 0, \mu \right),$$

$$\frac{dU_2}{dt} = A_2 U_2 + \mu \Pi_2 \left(U_1, U_2, \mu \right).$$

$$(137)$$

Критические переменные отщепляются в отдельную систему уравнений. Устойчивость нулевого решения системы (136) равносильны устойчивости нулевого решения первой системы (137).

Пример. Исследуем устойчивость нулевого решения системы дифференциальных уравнений

$$\frac{dZ_1}{dt} = \mu Z_2 + \mu^2 a Z_1^3, \quad \frac{dZ_2}{dt} = -Z_2 + \mu b Z_1^3. \quad (138)$$

Система уравнений (138) сводится к одному уравнению

$$\frac{dU_1}{dt} = \mu^2 (a+b) U_1^3 + O(\mu^3 U_1^4).$$

Поэтому система уравнений (138) имеет устойчивое нулевое решение при $a+b<0$ и неустойчивое при $a+b>0$.

Предложенный новый метод усреднения позволяет отщепить критические переменные.

Вопросы для самостоятельного решения

1. Рассмотреть функции от нескольких перестановочных матриц.

2. Разработать и опробовать разные численные методы определения ранга матрицы.

3. Разработать асимптотический метод в случае, когда матрица А в системе уравнений (119) имеет нулевые собственные числа.

4. Рассмотреть асимптотический метод в случае кратных собственных чисел матрицы А в системе уравнений (119).

5. Применить асимптотический метод к системам линейных дифференциальных уравнений с коэффициентами вида $\sum a_k e^{-\alpha_k t}$, $\operatorname{Re}\alpha_k > 0$.

Литература

1. Валеев К.Г. Расщепление спектра матриц. – Киев: Вища школа, 1986. – 272 с.

2. Гантмахер Ф.Р. Теория матриц. – М.: Наука, 1967. – 575 с.

3. Стрижак Т.Г. Асимптотический метод нормализации. – Киев: Вища школа,1984.–280 с.

4. Биркгоф Дж.Д. Динамические системы. – М.: Л.: ОНТИ, 1941. – 320 с.

5. Стрижак Т.Г. Метод усреднения в задачах механики. – Киев, Донецк, 1982. – 252 с.

ibidem-Verlag

Melchiorstr. 15

D-70439 Stuttgart

info@ibidem-verlag.de

www.ibidem-verlag.de
www.ibidem.eu
www.edition-noema.de
www.autorenbetreuung.de

www.ingramcontent.com/pod-product-compliance
Lightning Source LLC
Chambersburg PA
CBHW061200220326
41599CB00025B/4553